WEAVING SELF-EVIDENCE

PRINCETON STUDIES IN CULTURAL SOCIOLOGY

SERIES EDITORS: Paul J. DiMaggio, Michèle Lamont,
Robert J. Wuthnow, Viviana A. Zelizer

A list of titles in this series appears at the back of the book.

WEAVING SELF-EVIDENCE

A SOCIOLOGY OF LOGIC

CLAUDE ROSENTAL

Translated by Catherine Porter

PRINCETON UNIVERSITY PRESS

PRINCETON AND OXFORD

Published by Princeton University Press, 41 William Street, Princeton,
New Jersey 08540
In the United Kingdom: Princeton University Press,
6 Oxford Street, Woodstock, Oxfordshire OX20 1TW

Library of Congress Cataloging-in-Publication Data

Rosental, Claude.
[Trame de l'évidence. English]
Weaving self-evidence : a sociology of logic / Claude Rosental ;
translated by Catherine Porter.
p. cm. — (Princeton studies in cultural sociology)
Includes bibliographical references (p.) and index.
ISBN 978-0-691-13741-4 (hardcover : alk. paper)
ISBN 978-0-691-13940-1 (pbk. : alk. paper)
1. Logic, Symbolic and mathematical—Social aspects.
2. Reasoning—Social aspects. 3. Logic, Symbolic and mathematical—
History—20th century. I. Title.
Q175.32.R45T713 2008
511.3—dc22   2007049642

British Library Cataloging-in-Publication Data is available

Ouvrage publié avec le concours du Ministère français chargé de la culture–
Centre national du livre

The work is published with the cooperation of the Ministry of Culture–
National Book Center

This book has been composed in Sabon

Printed on acid-free paper. ∞

press.princeton.edu

Printed in the United States of America

10 9 8 7 6 5 4 3 2 1

CONTENTS

ACKNOWLEDGMENTS

A NUMBER OF PEOPLE and institutions have helped me carry out this project in many different ways. For their scientific and material support and for stimulating discussions, I am particularly indebted to the members of the Centre de Sociologie de l'Innovation at the École des Mines in Paris, the Science Studies Program and the Laboratory for Comparative Human Cognition at the University of California, San Diego, the Department of the History of Science at Harvard University, the Center for the Study of Language and Information at Stanford University, the School of Social Science at the Institute for Advanced Study in Princeton, and the Institut Marcel Mauss at the Centre National de la Recherche Scientifique in France. I owe much in particular to Michel Callon, Sabine Chalvon-Demersay, Clifford Geertz, François Jacq, Bruno Latour, Michael Lynch, Everett Mendelsohn, Chandra Mukerji, Louis Quéré, Sam Schweber, Joan Scott, and Brian Smith. I am also grateful to several colleagues whom I was fortunate to meet in various other institutions, including Éric Brian, Jean Louis Fabiani, Yves Gingras, Bernard Lahire, Michele Lamont, Oron Shagrir, and Gila Sher. Moreover, I would like to thank all those who have facilitated my empirical investigations, and especially Charles Elkan for making this research possible. I am also grateful to Catherine Porter for her wonderful work of translation and to Princeton University Press for publishing this book. Finally, I would like to express my deepest gratitude to the members of my family for their unstinting support.

WEAVING SELF-EVIDENCE

INTRODUCTION

IN A TEXT FROM the mid-1970s, a mathematician estimated that approximately one million theorems were produced in the world every five years.[1] If this affirmation still applies today, a whole set of questions arises at once. If we are indeed dealing with the authentic production of a massive number of statements, what "recognition" can a given theorem acquire, and to what reality does such recognition correspond? Through what pathways can a mathematical result—and its author, or even its challengers—obtain some form of accreditation? In other words, how does the collective validation of a theorem come about in practice? In particular, in fields that involve highly specialized skills, can the decisive role of "experts" be observed, and, if so, in what ways do the latter intervene in the process of certifying (or rejecting) a given theorem?

These questions will be addressed here on the basis of an actual case study documenting the way a theorem in mathematical logic was developed and accredited in the first half of the 1990s. On the basis of my own observations, textual analyses, and interviews, an empirical study carried out simultaneously with the logical research in question, I shall focus on the various tests to which this statement has been subjected and try to specify the factors on which the recognition it has achieved depends.

In so doing, I shall be exploring an avenue of research that has been largely neglected up to now by the social sciences. Indeed, while Émile Durkheim took "science" and especially "logic" as objects of sociological investigation par excellence,[2] the observations he set forth in 1912 concerning the unfortunate lack of pertinent empirical studies devoted to "logic"—studies destined to establish sociological analysis properly speaking[3]—still apply today. Without celebrating Durkheim's representations of the objects of logic and sociology, the present study aims to help make up for the lack of studies and analyses by seeking to account for certain observable modes of production of certified knowledge in logic in the early 1990s. My aim here is thus as much to take a step forward by offering for sociological reflection a "new" object of empirical investigation—logic being generally understood in the social sciences as an object essentially having to do with method—as to develop tools to describe this object.

---

[1] See in particular Stanislaw M. Ulam, *Adventures of a Mathematician* (New York: Scribner, 1976), p. 288.

[2] See Émile Durkheim, *The Elementary Forms of Religious Life*, trans. Joseph Ward Swain, 2nd ed. (London: George Allen and Unwin, 1976), pp. 437–38.

[3] Ibid., pp. 431–32.

But what is meant by *certified knowledge in logic*? One of the goals of this study is precisely to identify as specifically as possible the reality to which such a proposition may correspond. Thus I shall be able to attribute a precise meaning to this notion only at the conclusion of my work. However, I shall use the term provisionally here to refer to any statement that is relatively stabilized, more or less broadly accepted, advanced and validated by groups of actors in logic, whether they present themselves as philosophers, mathematicians, researchers in artificial intelligence, or in some other guise. I shall relate anything that these researchers may advance as "theorems" in mathematical logic, as philosophical statements in logic, or as statements attributing the working of a given system (especially in computer science) to logical formalisms.[4]

The problem raised in this book has parallels, needless to say, in various projects undertaken in the humanities and in philosophy, and especially in a set of analyses produced by sociologists of the so-called experimental sciences, analyses based on the study of the research process, the elaboration of facts, and the administration of proofs.[5] In the first chapter, I embark on a discussion with the authors of a certain number of these texts, in a dialogue that will continue throughout the book. For now, it seems important to clarify certain aspects of my approach to the project and of the empirical investigations I undertook in order to "observe" modes of production of certified knowledge in logic.

A SOCIOLOGIST AMONG LOGICIANS

I carried out my first observations during the academic year 1991–92 at the University of California, San Diego (UCSD). I first sought to identify as exhaustively as possible the various forms of logical activity that were being undertaken on that campus. Because I had been unable to find any sociological studies devoted to work in logic or to the profession of logi-

---

[4] The term "(logical) formalism" is used here to characterize a (logical) theory expressed with the help of a formal language. More detailed explanations of the various meanings given by me and other authors to this expression will be provided in the course of my analysis.

[5] On this last point, see especially Lorraine Daston, "Marvelous Facts and Miraculous Evidence in Early Modern Europe," *Critical Inquiry* 19 (1991): 93–124; Daston, "Objectivity and the Escape from Perspective," *Social Studies of Science* 2 (1992): 597–618; Simon Schaffer, "Universities, Instrument Shops and Demonstration Devices in 1776," Paris, Séminaire CRHST, "Spaces of Experiment," no. 2, May 10, 1994; Christian Licoppe, *La formation de la pratique scientifique: Le discours de l'expérience en France et en Angleterre (1630–1820)* (Paris: La Découverte, 1996); Alberto Cambrosio, Peter Keating, Thomas Schlich, and George Weisz, "Regulatory Objectivity and the Generation and Management of Evidence in Medicine." *Social Science and Medicine* 63 (2006): 189–99.

cian, the establishment of a cartography of logical activity seemed to be a necessary first step in order to arrive at an empirical understanding of the various exercises to which this term was applied, and in order to make initial contact with the actors involved, with an eye to later, more focused studies. Since there was no consensus among the faculty members I encountered as to how to define the field of logic—indeed, the matter often remained an object of debate—I studied the distribution of often mutually exclusive definitions, without making any a priori exclusions of my own. A phase of preliminary investigation, carried out through a series of interviews, analyses of written accounts of activities, and observations made during seminars, led me to a number of departments specializing in mathematics, philosophy, cognitive science, and computer science.

Prior to taking on this project, I had acquired a general background in science while I was taking more specialized courses in sociology, logic, history, and the philosophy of science. This itinerary turned out to be an asset that made it easy for me to engage researchers in dialogue and read their texts rapidly, without needing to begin by identifying and acquiring a mass of highly sedimented skills and knowledge. But training in mathematics and logic alone would have been insufficient for this project. It would have been difficult for me to construct a problematics from a sociological perspective, to formulate an empirical research project on such an object and not deviate from it, had I not had the opportunity to become familiar with an already vast literature in the sociology of the experimental sciences, and had I not already been interested in the notion of self-evidence in logic and in the possibility of developing a sociological approach to the issue. Since this specific form of "openness" compensated for the "difficulties" deriving from my own form of competence in logic and in science, my background was ultimately beneficial in that it enabled me to *circulate* freely within the framework of these preliminary investigations.

After I had grasped certain phases of the formation of statements in mathematical logic during research seminars by observing the participants' discussions, their interventions at the blackboard, and the written traces they produced, I was finally led by the relatively limited scope of logical activity practiced at UCSD to look for more appropriate sites in which to pursue my problematic. On the basis of bibliographical research and the testimony of various informants, I decided to continue my investigations at the Massachusetts Institute of Technology and Harvard University in a first phase, and at Stanford University and the University of California, Berkeley, in a second phase. Stanford had been represented to me on numerous occasions as one of the foremost centers of logical activity in the world.

Despite difficulties in setting up meetings at these sites with the faculty members involved and in obtaining authorization to observe certain of their activities, I was able to carry out a large number of interviews, take notes on interactions as they occurred, and observe and analyze a substantial production of texts (on paper, blackboards, or computer screens), especially during research seminars and lectures given by philosophers, researchers in artificial intelligence, and specialists in mathematical logic. In all these spaces, the production of logical knowledge was the work of scholars associated with a variety of disciplines: mathematics, philosophy, computer science, cognitive science, or linguistics.

During my study, I followed the day-by-day progress of several projects involving the simultaneous development of software and logical formalisms, projects carried out by teams in laboratories where I stationed myself as an observer. I used methods that had already been developed and tested in the sociology of the experimental sciences, bringing to bear ethnographic or ethnomethodological approaches.[6] Among the various forms of logic developed by the actors in question, I chose to follow systematically, in the various sites, the steps involved in developing and setting forth a specific logical formalism known as fuzzy logic. This decision was based primarily on a convergence of oral and written accounts that stressed the "tremendous flowering" of fuzzy logic at the time.

What were the material results of this course of action? All in all, between 1991 and 1996 I devoted about a year and a half to on-site observations. Supplemented by retranscriptions of interviews, these observations filled several notebooks. Some of the interviews, seminars, and project meetings were recorded on audio cassettes. I was also given permission to copy video recordings of certain project presentations and seminars, whether I had been in attendance or not. In addition, my research led me to collect a very large set of documents both on paper and in electronic form.

The process of reviewing and partially analyzing these data was spread over some three years. But since I have been talking about "observations," it is perhaps time to specify what sorts of things I was able to *see*.

OBSERVING DEMONSTRATIONS

In the course of my investigations, the actors I met characterized the objects of logic in contrary ways, sometimes arguing about them among themselves. Their own efforts were in fact invested in quite a wide range

---

[6] See especially Bruno Latour and Steve Woolgar, *Laboratory Life: The Social Construction of Scientific Facts* (Beverly Hills: Sage, 1979), and Michael Lynch, *Art and Artifact in Laboratory Science: A Study of Shop Work and Shop Talk in a Research Laboratory* (London: Routledge and Kegan Paul, 1985).

of activities. For my part, not wanting to prejudge what might be characteristic of their practices, I continually asked myself what it was important to observe. I sought to remain attentive to aspects that might have escaped me earlier, and I had many doubts about the object and the appropriate sites for my inquiry. Nevertheless, I was particularly struck early on by the considerable resources in time and energy deployed by the researchers in the activity of *demonstrating*. This work corresponded to the implementation of variable practices, as the following examples will make clear.

Let us consider first of all a project I followed in a laboratory at MIT involving the simultaneous development of a logical formalism and of software. Both were designed to be used to archive and annotate multimedia documents. This project brought together five researchers who spent most of their time preparing or carrying out, on the one hand, presentations at the blackboard among themselves or in front of others to explain the "principles" behind the software and the formalism to be implemented and, on the other hand, computer-based "demos" in front of small groups of invited guests, sometimes including the project's sponsors, to show them how the software worked. These activities were *constitutive* of the emergence of the formalism and the software, as much from the standpoint of pursuing and obtaining financing as from that of its content: indeed, the critiques formulated during these meetings largely determined the project's evolution, down to its smallest details. The moments of demonstration were *essential* to the advancement of this process, and in fact they structured the overall activity of the group.

In the same way, the time devoted to demonstration (including all the requisite preparatory steps) by academics affiliated with the departments of philosophy or mathematics at Stanford was also quite considerable. The preparation and presentation of articles or exposés at the blackboard, implying a large number of interactions and a material work of *writing*, played a central role in the formation and reformulation of statements or proofs. Moreover, the written or oral demonstrations gave rise to stagings that were as theatrical, as routine and stabilized, relatively speaking, as were the "demos."

The activity of logical production thus bore little resemblance to the image of actors working in near-total isolation, and it was not exclusively circumscribed by the minds of individuals. As a sociologist, then, I was not condemned to abandon my investigation for lack of competence to account for the activity in question. The work of logic unfolded through a great number of interactions, and these played an essential role in the production and transformation of statements. The activity involved also mobilized a huge gamut of material resources (written documents and arrangements for writing: blackboards, computers, laboratories, and so on), and its stakes varied according to the actors. In other words, logical activity looked like an authentic object for sociology, for it was eminently "social," if only because it involved groups of actors whose ties were in part forged or dissolved in the course of frequent and intense interactions.

Moreover, this activity did not appear "abstract" *in itself*: the process of abstraction could be grasped in its materiality.[7] While the activity had to be characterized as "theoretical," it seemed to me that that term should not be used in opposition to the term "practical," insofar as it implied a material work of writing that was carried out at various work sites and that called on a set of particular manual and visual skills. The expression "theoretical practice" thus seemed well suited to account for the specificity of the practices at work.

In other words, I found myself confronted with a large number of *mediations* of logic. I borrow the term *mediation* from studies in the sociology of music dealing with the way music is presented through the media.[8] The history of music seems in fact to have encountered the same difficulties with so-called immaterial objects as the history of logic and mathematics. I use the term "mediation," as opposed to "intermediary," in order to characterize all the resources that can be considered as "go-betweens" or conveyors (texts conveying ideas to readers or instruments conveying music to listeners, for example), and in order to make them visible as beings in their own right and as beings constituting the objects they are said to convey. During my investigation, texts and devices such as computers were often presented by researchers as putting the "principles" of a certain logic to work, while for me, as an analyst, they constituted mediations of that very logic.

Finally, I must stress the fact that the exercise of demonstration always struck me as the activity that best characterized the actors' work, structuring rather than punctuating it. The word "demonstration" implied variable practices—"demos," written proofs, and so on—that could be distinguished through observation, even if these practices were conflated under

---

[7] See Bruno Latour, "Sur la pratique des théoriciens," in *Savoirs théoriques et savoirs d'action*, ed. J.-M. Barbier (Paris: Presses Universitaires de France, 1996), pp. 131–46; for convergences with the program proposed by Éric Brian for a material history of abstraction, see Éric Brian, "Le livre des sciences est-il écrit dans la langue des historiens?" in *Les formes de l'expérience—une autre histoire sociale*, ed. Bernard Lepetit (Paris: Albin Michel, 1995), pp. 85–98. For a discussion of the place to be attributed to a materialist approach in the social and cognitive sciences, see Dan Sperber, "Les sciences cognitives, les sciences sociales et le matérialisme," in *Introduction aux sciences cognitives*, ed. Daniel Andler (Paris: Gallimard, 1992), pp. 397–420. See also the research in ethnomethodology on the modes of scientific interaction and production with the help of graphic constructions, especially on the basis of "following" a group of researchers in physics in Elinor Ochs, Sally Jacoby, and Patrick Gonzales, "Interpretive Journeys: How Physicists Talk and Travel through Graphic Space," *Configurations* 2, no. 1 (1994): 151–71.

[8] See Antoine Hennion, *La passion musicale* (Paris: Métailié, 1993).

a single umbrella term. The examination of modes of demonstration focused primarily on the intersection between the analysis of technologies of proof and the analysis of forms of ostentation. The observation of a wide range of demonstrative practices, with significant corresponding divergences of principle, allowed me in any event to grasp, without having to pass judgment as to the legitimacy of any particular point of view, a remarkably "chaotic" situation: statements that certain specialists took to be "rock solid" were contested just as vigorously by others, so that there was, on the whole, a pronounced splintering of the logical certainties advanced.

A NECESSARILY DETAILED ANALYTIC ACCOUNT

In the face of such phenomena, the existing models of the nature and methods of logic that had been developed by the philosophy of logic struck me in numerous instances as of quite limited usefulness. Once the meticulous work of observing had been completed, the task of describing and analyzing proved very complex. Indeed, I found no way to avoid fine-grained descriptions if I wanted to account for the dynamics I had observed in terms understandable to nonspecialists. I could not simply refer to specific documents and developments; I had to present them in their "raw" state. The constraints had to do in particular with the large number of specialized practices and skills brought into play and their deep sedimentation: up to that point, their circulation had been limited to highly restricted circles, which meant that they would be virtually inaccessible to almost all readers unless special efforts were made on my part—all the more so in that by and large philosophy had not taken up these practices and skills as objects of study and thereby done the work of mediation.

In the face of this last difficulty, highlighting a particular dimension of logical activity and the specificity of the objects of the philosophy of logic, I finally decided to focus my analysis initially on a specific type of statement, namely, *theorems* in mathematical logic. Given that theorems are the objects generally perceived (rightly or wrongly) as being at the heart of what logicians do, I chose to study the way they emerge and the way they achieve the status of certified knowledge.

It soon became clear that if I wanted to account in a satisfactory way for the emergence of even one theorem, an entire book would barely suffice. This observation as such represents an important result of my research. This book thus constitutes the product of a necessary, drastic selection from among a very large set of data bearing on the work of actors in logic.

How was I able to settle on one particular theorem, given an annual production that appears quite colossal on a worldwide scale? My choice corresponded first of all to an exceptional opportunity to follow the various phases of the production and presentation of a logical theorem on a "relevant" time scale (the meaning of "relevance" here will become fully clear at the end of this study). The choice also corresponded to a certain "convenience" factor: It would give my readers access to an excellent observation point for a certain number of demonstrative practices and modes of production of certified knowledge in logic, which I was able to find at other sites of investigation as well. In this sense, my choice can be compared to the geneticists' choice of the fruit fly. The metaphor has its limits, however, given that my own inquiry was not based on performing experiments, and of course I do not presuppose the "universality" of certain results.

To be sure, I went beyond the use of methods that some might describe as microsociological and sought insofar as possible to circumscribe the scope of the practices described, to grasp their historical emergence, and to develop a cartography of the relations and groups involved. However, since there was not an already-constituted body of sociological or sociohistorical knowledge of logic on which I could rely, I was unable to go as far in this undertaking as I would have liked; I have only been able to sketch its broad outlines.

In this context, if my research allows glimpses of other pertinent avenues for research, if it offers useful tools for investigation and shows the value of in-depth study of practices and strategies that have been overlooked in sociology and social history, although they correspond to a dimension of scientific activity that is often presented as crucial, it will already have achieved a nonnegligible goal. I should like to think that it might go even further and contribute to a renewal of the questions raised or even of the models presented in the philosophy of knowledge that have to do with the methods and nature of logic.

GRASPING THE EMERGENCE OF A LOGICAL THEOREM

The purpose of the analysis that follows is thus, as promised, to attempt to account for the emergence of a particular logical theorem at the beginning of the 1990s and to study to what extent and by what means it managed to achieve the status of certified knowledge. The first question to which we must return is methodological in nature: how was I able to have access to such a process? The description of my itinerary will give a first glimpse of certain logical practices that I shall analyze later in more detail.

As I mentioned earlier, I found myself on the University of California, San Diego campus during the 1991–92 academic year, and I tried to sort out as exhaustively as possible the various forms of logical activity that were carried out in that setting. This led me to a number of different academic departments; a member of the cognitive science department suggested that I should talk to Charles Elkan, an assistant professor of computer science.

During our conversations, Elkan stressed the importance of research on artificial intelligence in computer science departments in the United States. He pointed out the existence of "trends" and remarked on the great diversity of pro-logician and anti-logicist approaches in artificial intelligence. In his descriptions, intellectual and professional investments were closely intertwined. Elkan displayed a pronounced preference for one particular school of thought, indicating his particular interest in the "classical" logical theory known as "first-order logic" and in an "important" logical theory in artificial intelligence in the United States known as "nonmonotonic logic." He cited a certain number of pioneers in this field who were working primarily in the San Francisco Bay area, at Stanford and Berkeley.

According to him, the relatively recent developments of fuzzy logic and of a theory in artificial intelligence called "neural networks"[9] used much poorer and less effective tools than those of "classical" artificial intelligence. When I asked Elkan on what this judgment was based, he said that he had formed his first opinions about fuzzy logic by reading some articles in the mass media about its basic principles and applications to household appliances. To become better informed, he had read an article published in a scholarly journal by an author cited as one of the principal specialists in fuzzy logic.[10] He had also participated in a conference in Australia that brought together researchers working on the "applications" of fuzzy logic.

The definitions of fuzzy logic and the demonstrations of its properties to which Elkan had been exposed led him to formulate a rather negative judgment about the theory. He asserted that the "applications" functioned well, but that this functioning should not be attributed to fuzzy

[9] The name refers to a method that seeks to exploit and implement artificially a biological process attributed to the brain and presumed to account for intelligence: interconnection of neurons in networks. See P. M. Churchland, *A Neurocomputational Perspective: The Nature of Mind and the Structure of Science* (Cambridge: MIT Press, 1989). See also J. D. Cowan and H. D. Sharp, "Neural Nets and Artificial Intelligence," in *The Artificial Intelligence Debate: False Starts, Real Foundations*, ed. S. R. Graubard (Cambridge: MIT Press, 1989), pp. 85–121.

[10] Bart Kosko, "Fuzziness vs. Probability," *International Journal of General Systems* 17 (1990): 211–40.

logic as such. He further declared that he planned to write an article in which he would offer proof that fuzzy logic was contradictory, and he sent me the paper in draft form.

Several months later, early in the summer of 1993, Elkan sent me another version of his article, which was to be presented in July and published in the proceedings of the annual conference on artificial intelligence in the United States, AAAI'93. In this text, "The Paradoxical Success of Fuzzy Logic,"[11] the author undertook a critical analysis of both the "practical" and the "theoretical" developments of fuzzy logic. He began with an attempt to offer proof that a version of fuzzy logic characterized as "standard" and described as a formal system endowed with a specific axiomatics was "in fact" only ordinary logic with two truth values (true and false), even though it was generally presented—still according to the author—as a logical system allowing an indefinite number of truth values. The axioms that constituted the point of departure of this demonstration were four in number (here, "∧" represents the logical connector "and," "∨" the connector "or," "⌐" the connector "not," and "t" represents the truth value of the assertion in parentheses:

$$t(A \wedge B) = \min \{t(A), t(B)\}$$
$$t(A \vee B) = \max \{t(A), t(B)\}$$
$$t(\neg A) = 1 - t(A)$$
$$t(A) = t(B) \text{ if } A \text{ and } B \text{ are}$$
logically equivalent.

In support of the path his demonstration had followed, Elkan stated the following theorem: "For any two assertions A and B, either $t(B) = t(A)$, or $t(B) = 1 - t(A)$." Elkan "immediately" deduced from this that fuzzy logic, as a formal system characterized by the four preceding axioms, was "in fact" a logical system with two truth values (zero and one).[12]

Elkan's formulation and his line of reasoning already raise a large number of questions on several levels. At this stage, however, my purpose is not to explore these questions. The access that I seek to construct for my readers, corresponding in particular to the explicitation of the resources that a theorem like Elkan's mobilizes, is a principal concern in the developments to follow. The same thing is true of the proof, which I shall analyze in detail later on.

[11] Charles Elkan, "The Paradoxical Success of Fuzzy Logic," *Proceedings of the Eleventh National Conference on Artificial Intelligence* (Menlo Park, Calif.: AAAI Press; Cambridge: MIT Press, Aug. 1993), pp. 698–703.
[12] Ibid., p. 698.

Following this demonstration, Elkan set forth some critical developments regarding the "applications" of fuzzy logic. The following theses were announced in the summary that preceded his article:

Fuzzy logic is not adequate for reasoning about uncertain evidence in expert systems.[13] Nevertheless, applications of fuzzy logic in heuristic control[14] have been highly successful. We argue that the inconsistencies of fuzzy logic have not been harmful in practice because current fuzzy controllers are far simpler than other knowledge-based systems. In the future, the technical limitations of fuzzy logic can be expected to become important in practice, and work on fuzzy controllers will also encounter several problems of scale already known for other knowledge-based systems.[15]

Elkan's logical proof lay at the heart of the arguments he developed in his article. This was also the case for the structure of the major controversy, involving numerous researchers in artificial intelligence, that his text provoked as soon as it appeared. The controversy led the author to produce new written and oral formulations of his theorem and its proof, and to take many accompanying actions (the exact nature of what I am calling "accompanying actions" will become clear in due course). This, then, is how I finally found myself confronting phenomena related to the development of a logical theorem, phenomena for which I now propose to account.

What is the best way to grasp these debates and the various undertakings of the participants? While the emergence of a logical theorem, the characterization and modalities of its access to the status of certified knowledge, had never before been the object of a sociological investigation, to my knowledge,[16] certain methodological operations used in soci-

---

[13] The expression "expert system" has largely become synonymous with "application of artificial intelligence," inasmuch as it is a system of rules making it possible to carry out a "reasoning" process through a computer (for example, to diagnose a machine breakdown). See P. S. Sell, *Expert Systems: A Practical Introduction* (New York: John Wiley, 1985), and Frank Puppe, *Systematic Introduction to Expert Systems: Knowledge Representations and Problem Solving Methods* (New York: Springer-Verlag, 1993).

[14] The author is referring here to all the little "automatic" control devices that are included, for example, in household appliances. On this topic, see for example L. A. Zadeh, "Fuzzy Logic," Stanford University, CSLI Report no. CSLK-88–116 (1988): 20–25. See also R. N. Clark, *Introduction to Automatic Control Systems* (New York: John Wiley, 1962), and Greg Knowles, *An Introduction to Applied Optimal Control* (New York: Academic Press, 1981).

[15] Elkan, "Paradoxical Success," August 1993, p. 698.

[16] As we shall see, Donald MacKenzie's study of the debates over the formal verification of a microprocessor nevertheless offered some interesting points of comparison for my research. See MacKenzie, "Negotiating Arithmetic, Constructing Proof: The Sociology of Mathematics and Information Technology," *Social Studies of Science* 23 (1993): 37–65.

ology to study the experimental sciences could nevertheless be drawn upon profitably here to apprehend such a dynamic. To "follow" the controversy, I looked closely for written traces in various media, and I completed the study through a series of interviews. Oral testimony and written references to other texts were the key elements that allowed this step-by-step approach. However, if it is the case, as I shall attempt to show, that phenomena associated with "referring practices" were integral to the demonstrative strategies that I shall have to make explicit, we can then consider that the principle of circulation was supplied by the various demonstrations of the actors themselves. Thus I developed my own methods of investigation by taking into account the specificity of the set of practices observed—for these practices were quite different from the ones encountered in studies focusing on the experimental sciences.

Moreover, from the standpoint of the skills characteristic of a sociologist, this empirical undertaking could have been relatively classical if, among the media constituting the framework in which the controversy unfolded, there had not been an electronic forum on the Internet. As we shall see, this forum played a very important role in the observation of certain phenomena of which it is in part the origin.

I had discovered the existence of this forum in early 1994 during a conversation with a researcher in fuzzy logic. Numerous messages related to the controversy had been exchanged previously and were no longer directly accessible, but, fortunately, electronic archives had been created. With the necessary computer resources at hand, I was able to access the archives, carry out the corresponding work of connection, collection, and analysis (a new task for a historian dealing with recent events), and attempt to reconstitute the evolution of the debates.

Accessing the Specialized Skills of Workers in Logic

In the hope of enabling readers to appreciate the intricacies of this process, to assess the progressive and relative recognition of the theorem (or rather of the successive theorems), and to grasp the operators of stabilization and the demonstrative practices at work in the debates, I have adopted a particular narrative structure. While I referred earlier to my intention to try to provide readers with a means of *access* to Elkan's theorem, my project in fact extends to the dynamics of production of certified knowledge in its entirety.

This approach entails determination on my part to surmount one of the major difficulties typically faced by a sociological approach to such an object: the risk of producing a text that the bulk of the potential audience, including students and researchers in the social sciences, will find unreadable. To avoid ending up with a work that is accessible solely to those who practice the activity described, it was important at the outset not to resort to a descriptive vocabulary that would simply echo that of the actors (for want of interrogating it adequately). In reality, I had to take at least two types of readers into account.

First of all, I wanted to provide readers who might have little or no knowledge of logic, and who might assume that such a study was inaccessible, with all the elements necessary for grasping the phenomena under consideration. To this end I traced back certain modalities for learning the practices involved, and I gradually constructed a vocabulary suitable for describing them.

However, I did not want to overlook those who already had some background in logic, and who might be tempted to go "straight to the point," that is, to "form an opinion" about Elkan's demonstrations and those of other researchers, according to a well-internalized academic or professional custom, on the basis of their own practice of logic, without even noticing that the problematics of this research might lie elsewhere. The goal, after all, is to provide the means for analyzing the modalities of a dynamics of production of certified knowledge; thus "judging" the researchers encountered and their production, deciding that so-and-so is "right" or "wrong," is not in any sense the aim of this study.

A reader who formed an opinion at the outset and who deemed, for example, that a particular actor "was right" on the basis of that actor's interventions would still not have the means to know whether the latter had succeeded in rallying other participants to his or her point of view and had actually won the debates. At this stage, readers are actually in the same situation as the researcher was some time earlier: if they want to know what factors played a role in the evolution of the debate and its eventual outcome, they must not neglect any path a priori. That is why I shall spend time analyzing the exercise of logic in detail, by examining the modes of intervention of the various protagonists in the debates (and in particular the modalities through which they take the floor or, more often, intervene in writing).

In this sense, the itinerary proposed to the first type of reader is perfectly adapted to the second type as well. In particular, presenting and analyzing Elkan's proof only at the *end* of the process makes it possible, on the one hand, to supply the "naïve" reader with all the elements needed to grasp it, and, on the other hand, to keep the knowledgeable reader from jumping the gun and mistaking the nature of the exercise.

But let me be more precise about the steps in this process. The Internet discussion group to which I have referred was an essential forum, as we shall see: the one in which the first "public" reactions to Elkan's article appeared, owing to the lag in publication of scholarly journals; I shall thus begin by examining the noteworthy exchanges that appeared there. Before doing so, I shall take care both to offer the necessary details about the way the forum works and also to use the context of a classroom to analyze the development of the logical vocabulary and the techniques that a number of the actors share (chapter 2). I shall then plunge directly into the heart of very diverse logical practices and into literally cacophonic interventions that will warrant in-depth analysis (chapters 3 and 4).

We shall then look at other mediations of the debates, in particular the various actions that accompany the proofs, publications in other media, and assorted reformulations of the demonstrations, from the very first drafts of Elkan's article to the publication of diverse versions of the text. In this way I shall try to account in a detailed way for the process of producing certified knowledge, for the dynamics of the controversy and the mode in which it reached a stable state. In particular, we shall attempt to understand how, beginning from so many different viewpoints and divergent approaches to the practice of logic, certain researchers succeeded in occupying center stage and certain overtly federating positions managed to impose themselves (chapters 5, 6, and 7).

However, before embarking on such an analysis, it is essential to note the contributions of particular works to the design of this study, along with the questions that these works raise. As a first step (chapter 1), an examination of specific approaches to logic and mathematics in the human sciences and in philosophy will enable us to identify perspectives from which one can gain *access* to the highly closed world of actors in logic.

PART ONE

ACCESSING THE WORLD OF PRODUCERS OF LOGICAL STATEMENTS

What has been able to make social life so important a source for the logical life? It seems as though nothing could have predestined it to this role, for it certainly was not to satisfy their speculative needs that men associated themselves together. Perhaps we shall be found over bold in attempting so complex a question here. To treat it as it should be treated, the sociological conditions of knowledge should be known much better than they actually are; we are only beginning to catch glimpses of some of them.
   —Émile Durkheim, *The Elementary Forms of Religious Life*,
      trans. Joseph Ward Swain, 2nd ed. (London: George Allen & Unwin, 1976), p. 432

Chapter 1

HOW CAN WE GRASP WHAT LOGIC-MAKERS DO? QUESTIONS RAISED IN THE HUMAN SCIENCES AND PHILOSOPHY ABOUT LOGIC AND MATHEMATICS

WORKS THAT HAVE some relevance to our problematic, works whose analysis may offer modes of *access* to the field I have taken as my object, clearly constitute a very large set. Within this set, a certain number of empirical approaches to logic and mathematics, scattered throughout an abundant and heterogeneous literature, are of particular interest for the investigation to be pursued. These approaches contribute specifically to the process of negotiating[1] definitions of sociology, history, anthropology, ethnology, psychology, and philosophy—but also and above all of logic and mathematics—as they are practiced. Neither a history of ideas, no matter how nuanced, nor a compartmentalized disciplinary history would be at all adequate for the task of accurately representing the sides taken and the positions articulated in the debates spawned by these negotiations. Without falling back on the history of ideas, then, but also without launching into a description deriving from an approach based on social history (the space of a single book would hardly suffice for this), it seems indispensable to select and present certain direct and indirect representations of the objects to be studied in order to lay out the problematics and my analysis on the basis of the questions these representations raise. I shall thus be able to specify the paths it will be appropriate to follow in order to approach the field of inquiry selected.

At a later stage, we shall need to take into account a series of recent studies in sociology and in the history of the experimental sciences that focus on the ways proofs are produced. At this point, my aim is simply to carry out a critical analysis of works that form a limited frame of reference and interaction, works that either use an *empirical* approach to logic and mathematics or, if they do not put such an approach into practice, at

---

[1] By negotiation I mean the positing of antagonistic and performative definitions (i.e., definitions that constitute their object through the very act of utterance) concerning the various research fields and their practices.

least advocate the principles of empirical inquiry into the concrete production of knowledge in logic.

In a first phase, I shall look at several ways in which the empirical study
of logic has been developed in the sociology of knowledge, in psychology,
and in the ethnology of reasoning, and show how they have all *started*
with an *a priori* and *stabilized* definition of the objects of logic, rather
than apprehending those objects in their dynamic dimension and as products of a process of certifying knowledge. While certain philosophers do
not rule out an empirical approach to logic, I shall show how the philosophy of Ludwig Wittgenstein marks a noteworthy departure insofar as the
importance granted to the *practices* of actors in logic or mathematics is
concerned. I shall then look at a certain number of works developed in
the history and sociology of mathematics in Wittgenstein's wake: for example, studies by David Bloor, Andrew Pickering, and Andrew Warwick.

Moving away from this tradition, we shall see how Éric Brian's sociology of mathematics, without being a disciplinary history, makes it possible to analyze mathematical activity in a convergent fashion, in particular
through the *practices* that this activity entails. I shall then pursue my critical analysis by focusing on some attempts—carried out largely by professional mathematicians—to describe contemporary mathematical work
from a sociological perspective. Last, I shall look at a few isolated case
studies in sociology and in the history of science whose authors highlight
logical practices or formulate a program for empirical investigation into
these practices.

DO RESEARCHERS IN LOGIC INVOKE IDEAL PRINCIPLES?

Logic as a mode of reasoning mobilized in various human activities (and
not specifically as an object of production of knowledge certified by scientists—the distinction is a crucial one) has given rise to a good deal of
empirical research since the end of the nineteenth century. Several authors
have seen it as a key object of investigation for the sociology of knowledge. This was true of Émile Durkheim, who took both science and logic
as objects of sociological investigation par excellence.[2] In Durkheim's
analysis, the religious phenomenon was at the origin of the fundamental
notions of science, and especially of logic.[3] To found sociological analysis
properly speaking, however, as we have seen, Durkheim called for thoroughgoing empirical investigations; the lack of such studies was a serious
problem in his eyes.

---

[2] Durkheim, *Elementary Forms of Religious Life*, p. 438.
[3] See ibid., pp. 431–32.

Lucien Lévy-Bruhl,[4] for his part, sought to show the connection between the more or less "logical" behavior of individuals and their status as members of "primitive" peoples or of Western societies. From this perspective, carrying out an ethnology of logic, so to speak, consisted in taking into account an a priori representation of the nature of logic and its immutable principles, and then studying the greater or lesser correspondence between these principles and the way individuals think, by linking the principles to certain individual behaviors (the process of establishing relationships constituted part of the analytic work properly speaking). Thus, for example, Lévy-Bruhl associated the prelogical "primitive mentality" with lack of respect for the principle of contradiction.[5] Lévy-Bruhl's problematic found echoes throughout the twentieth century, and even today we find ethnological works that reformulate similar questions on the basis of new empirical research.[6]

Similarly, it is important to note the existence of a no less ancient research problematics that consists in interrogating the psychological factors in reasoning that underlie the individual's respect or nonrespect for logical principles (such as the principle of contradiction). Efforts to interpret the results of numerous experiments carried out in this area have led to controversies among cognitive psychologists that remain quite heated.[7] In the context of these debates, a consensus often develops among analysts as to the nature of logic, which is generally perceived as stable and which is *identified* in practice with a small number of syllogisms. In contrast, analysts may disagree as to how effectively logic was used in a given argument, or why it was not brought to bear at all.

Thus, for example, starting from an account of a specific experiment, a polemic may develop as to whether a particular syllogism was actually

[4] For a comparison between Durkheim's theses and Lévy-Bruhl's and an analysis of the way these questions are dealt with, see the work of Robin Horton: Robin Horton and Ruth Finnegan, *Modes of Thought: Essays on Thinking in Western and Non-Western Societies* (London: Faber, 1973), and Robin Horton, "Tradition and Modernity Revisited," in *Rationality and Relativism*, ed. Martin Hollis and Steven Lukes (Oxford: Blackwell, 1982), 201–60.

[5] Lucien Lévy-Bruhl, *Primitive Mentality*, trans. Lilian A. Clare (Boston: Beacon Press, 1966 [1922]). On the evolution of Lévy-Bruhl's theses, see Jean Cazeneuve, *Lucien Lévy-Bruhl* (Paris: Presses Universitaires de France, 1963).

[6] See for example James F. Hamill, *Ethno-logic: The Anthropology of Human Reasoning* (Urbana: University of Illinois Press, 1990). See also, with its study of the Azande people and the way they apprehend contradictions, E. E. Evans-Pritchard, *Witchcraft, Oracles, and Magic among the Azande* (Oxford: Clarendon Press, 1937).

[7] See especially Vittorio Girotto, "Judgements of Deontic Relevance in Reasoning: A Reply to Jackson and Griggs," *Quarterly Journal of Experimental Psychology* 45A, no. 4 (1992): 547–74. See also Gerd Gigerentzer, "From Tools to Theories: A Heuristic of Discovery in Cognitive Psychology," *Psychological Review* 98, no. 2 (1991): 254–67.

brought into play. In such cases, it is not unusual for other modes of reasoning to be advanced and for their origin to become the object of debates in turn: attributed by certain authors to the "particular conditions of the experiment," this origin may be associated by others with "cultural factors," for example, or with a "social context," or with the "education of individuals."[8]

We shall not linger over this type of work, since the present project has a different focus. It aims to account for the specific dynamics of production of certified knowledge in logic. Thus it does not seem appropriate to propose a *stable, a priori* definition of logic in order to *confront* it later on with empirical investigations. Rather, it seems legitimate to try to determine whether "logic," in the sense in which certain ethnologists and psychologists understand it, is the *same* thing as what the researchers we shall encounter understand by logic. In particular, it is a matter of finding out whether, for these researchers, logic "rests" on the ideal necessity of principles such as the principle of contradiction, or whether it is the object of distinct conceptions and practices.

How Do Multiple Social Actors Put Forward Various Definitions and Practices of Logic?

The philosophies of logic that have been developed since the late nineteenth century have not all attributed the same value to carrying out empirical investigations of the process by which knowledge is produced in logic. Many philosophers simply failed to consider such an undertaking, while others viewed the prospect as of very limited interest. Still others saw it as irrelevant to, and outside the scope of, philosophical research.

Given the abundance of the literature involved and the considerable divergences between the problematics at stake in that literature and the one that underlies my own study, I do not propose to explore even a modest sampling here. At this stage I shall simply evoke a few specific representations, in order to suggest the degree to which the divergences mentioned are significant. However, I shall return to this question later on, when I take these problematics not as tools but as *objects* of analysis.

When Edmund Husserl introduced his phenomenological project, he insisted on the impossibility of resorting to psychological research, or to empirical research in general, to account for the "ideal foundations" of

---

[8] See for example the discussions in Guy Politzer, ed., "Pragmatique et psychologie du raisonnement," special issue, *Intellectica* 11 (1991).

logic.[9] He was referring to efforts by Christoph Sigwart[10] and Benno Erdmann,[11] among others, to establish psychological and anthropological
foundations for logic.

Sigwart and Erdmann had each produced a treatise developing his own
"anthropology" of logic, in the form of a discussion of principles whose
object was to show that the laws of logic were essentially tied to the nature
of human beings (as a species) and to human forms of thought.[12] It is
important to note that Husserl's opposition to these writers bore explicitly on the question of the "ideal foundations" of logic. As Husserl's
phenomenological project clearly showed, the "subject's" actual "access"
to self-evidence or to logical necessity was an entirely legitimate object
for research, in Husserl's view, so long as it fit into the categorization he
proposed.[13]

Similarly, the logician and philosopher of logic Rudolf Carnap distinguished the ideal autonomy of a system of logical syntax from the existence of a process by which "man" may gain "access" to this system.[14]
While he asserted that this process was not the object of his research, at
least the process itself was not called into question.

I shall provide only a minimal description here, since the tasks assigned
by these philosophers to an empirical approach to the production of
knowledge in logic is, as we can see from the outset, simply too far removed from the paths this book is seeking to explore. Its objective is
neither to carry out a preliminary discussion of principle concerning what
logic *is* nor to proceed to an empirical investigation concerning the way
"logic" may be *apprehended* by *a knowing subject*. In the context of this
undertaking, it is essential not to be too quick to singularize either the
objects of logic or the actors (who are to be contrasted with a stereotypical
"knowing subject"); on the contrary, it is critical to concentrate in a first

---

[9] See Edmund Husserl, *Logical Investigations*, trans. J. N. Findlay (London: Routledge,
2001 [1901]).

[10] Christoph Sigwart, *Logik* (Tübingen: H. Laupp, 1889).

[11] Benno Erdmann, *Logik* (Berlin: W. de Gruyter, 1923).

[12] It would be interesting to compare these works with the anthropological theses formulated much later by W.V.O. Quine, but such an analysis would exceed the scope of the
present work.

[13] It should be noted that a sociology of knowledge with a phenomenological bent, stressing the ideal character of knowledge but taking as its object the factors of the more or less
rapid emergence of a given "genre" of knowledge (mathematics, physics, and so on), was
developed by Max Scheler, in *Die Wissenformen und die Gesellschaft* (Leipzig: Der neue
Geist, 1926).

[14] See Rudolph Carnap, *The Logical Syntax of Language* (London: K. Paul, Trench,
Trubner, 1937 [1934]). The dual status I attribute to Carnap—that of logician and philosopher of logic—would warrant a thorough investigation in the framework of a social history
of twentieth-century logic.

phase on grasping the potential plurality of both objects and actors in logic on the basis of a concrete work of observation.[15]

Instead of contriving to *stabilize a priori* either the "subjects" or the "objects" of logic, it will be important to try to account for the modes through which groups of social actors (who remain to be identified) can be led to negotiate both the definition and the practice of these objects. Whereas Husserl sets forth logical principles such as the principle of contradiction and Carnap relies on syntactic foundations in attempting to "found" logic as a discipline, the project at hand includes analyzing the way the researchers we encounter mobilize divergent logical concepts (philosophical and otherwise) and practices.

This approach appears to be all the more necessary in that, given the extent to which logical activity is scattered among a variety of disciplines, I ended up conducting *observations* (and not discussions) that bore upon the work and the production of contemporary philosophers of logic. As we shall see, the production of philosophical considerations constitutes an inherent *resource* for the activity of researchers in logic, whether they are professional philosophers or not: such considerations can easily be mobilized—perhaps more easily than in other scientific domains—within the framework of their exchanges. Thus it seems quite simply inappropriate to *debate* such considerations within the framework of this project. In contrast, it will be necessary to *take them into account* as objects of analysis in their own right.[16]

[15] From this standpoint, see Brian Rotman, *Ad Infinitum—the Ghost in Turing's Machine. Taking God Out of Mathematics and Putting the Body Back In: An Essay in Corporeal Semiotics* (Stanford, Calif.: Stanford University Press, 1993). By introducing the triple body of "the" mathematician in the framework of his semiotic approach to mathematics, Rotman takes a first step toward making the mediations of mathematical activity and its actors more visible. This step now calls for others. It would be appropriate in particular to examine the extent to which the tripartite division of the mathematician's body that Rotman constructs in his philosophical essay is well defined uniformly and ideally by the set of mathematical texts produced throughout history. Would such a tableau offer a not too costly way of circulating within the history and sites of logical and mathematical practices? Must we posit a univocity of mathematical signs, or on the contrary sometimes expect the failure of effects that seem the most highly programmed in them? To answer such questions, it seems indispensable to examine what may be a multiplicity of readings of mathematical texts, corresponding to practices that may be situated historically and geographically or may be more or less spread out in networks. The study that follows, in which demonstrations will be confronted with radically divergent modes of reading and writing deployed by situated researchers endowed with specific competencies, will allow me to contribute some partial responses to these questions. For other semiotic studies of mathematics, see also Brian Rotman, *Signifying Nothing: The Semiotics of Zero* (London: Macmillan, 1987), and Edwin Coleman, "The Role of Notation in Mathematics," Ph.D. diss., University of Adelaide, Australia, 1988.

[16] See also the approach taken, in the context of the development of a social history of philosophy, in Jean-Louis Fabiani, *Les philosophes de la République* (Paris: Minuit, 1988).

In the process I shall try to grasp the way various philosophies of logic, introduced or reformatted by the actors, correspond in fact to specific logical practices and play a role in the mobilization of these practices. By bringing to light the multiplicity of skills and options adopted in logic, a complex corresponding as a whole to an amalgam among competing philosophies of logic, I shall be able to make sense *a posteriori* of a singular phenomenon: the abundance of declarations made by practitioners of logic denouncing misunderstandings that are alleged to bear upon the nature of logical activity and its objects, and of which numerous philosophers are said to be victims.

These denunciations, which practitioners formulate when they perceive a significant disparity between their own activity and the activity that certain philosophies of logic attribute to them, may take various forms. As an illustration, let us look at two excerpts from denunciations that have been given *public* expression, in writing:

Some books bear the title *Philosophy of Logic* . . . Full of useful information, they leave an ambiguous impression: either the heart of the matter has escaped them, or else logic has no more import than the art of playing chess. . . . In our day, professional philosophers overestimate logic: against all evidence, they believe that mathematicians use it. . . . The patronage of logic also lends the prestige of science to speculations about the theory of knowledge, it being understood that the mark of science is a calculation or a rule. Mechanics, or what can be made mechanical, would be equivalent to perfect discursivity, to pure rationality. . . . Earlier, mathematicians used to warn against this sort of error. . . . They have not stopped conforming to these principles in their daily work; more recently, some have used a different discourse, and they bear some responsibility for the stupidities of epistemology. . . . The logic of mathematical language has been exploited in turn by dubious epistemologies . . . , where it had no role to play.[17]

The fake philosophical terminology of mathematical logic has misled philosophers into believing that mathematical logic deals with truth in the philosophical sense. But this is a mistake. Mathematical logic does not deal with truth, but only with the game of truth. The snobbish symbol-dropping one finds nowadays in philosophical papers raises eyebrows among mathematicians. It is as if you were at the grocery store and you saw someone trying to pay his bill with Monopoly money.[18]

[17] Jean Largeault, *La logique* (Paris: Presses Universitaires de France, 1993), pp. 122–25.
[18] Gian-Carlo Rota, "The Pernicious Influence of Mathematics upon Philosophy," *Synthese* 88 (1991): 168. See also Jacques Bouveresse, *Prodiges et vertiges de l'analogie: De l'abus des belles-lettres dans la pensée* (Paris: Raisons d'agir, 1999), pp. 39–40.

Later on, we shall be able to look closely at the objects of these denunciations, which are clearly embedded within very rich conflictual configurations, and I shall attempt to make them at least partially accessible to the reader. For now, I want to emphasize the quite specific interest that Wittgenstein's work offers for our problematics.[19]

Indeed, few philosophers have done as much as Wittgenstein to stimulate research in the social sciences on logic and mathematics. In analyzing logic and mathematics through the practices they involve and the activity they represent (particularly through studies of language games), Wittgenstein has been a reference for a number of contemporary studies in sociology and the history of science, and especially in the social history of mathematics. Certain works in this latter field show in turn that it is possible to go beyond a unitary and often allusive demonstration and bring to light the existence of mathematical practices that are historically situated, more or less long-lived, localized or mobilized in networks.

This is the path I shall try to follow, given the logical practices that I have observed and studied. In order to develop useful tools for questioning, I now need to introduce some of the above-mentioned works in the history of mathematics.

QUESTIONS RAISED BY CERTAIN WORKS IN THE SOCIAL HISTORY OF MATHEMATICS

Can Institutional Sociology Account for the Ways in Which Research in Mathematics Is Carried Out?

In the mid-1970s, David Bloor drew upon a series of Durkheimian concepts to formulate a research program in the sociology of logic and mathematics, putting it to work on several historical cases.[20] In Wittgenstein's writing, Bloor finds a philosophy in accord with the Durkheimian sociology of logic that he is seeking to develop. According to Bloor, Wittgenstein succeeded in showing that logical necessity is of the same order as moral necessity: we are constrained by logic as we are constrained to accept the well-foundedness of certain behaviors, because we take a certain way of living for granted.[21] Bloor thus views as ill-founded Karl Mannheim's the-

---

[19] See Ludwig Wittgenstein, *Philosophical Investigations*, trans. G.E.M. Anscombe (Oxford: Blackwell, 1953), and Wittgenstein, *Remarks on the Foundations of Mathematics*, trans. G.E.M. Anscombe (Oxford: Blackwell, 1956).

[20] David Bloor, *Knowledge and Social Imagery* (Chicago: University of Chicago Press, 1991; 1st ed. London: Routledge and K. Paul, 1976).

[21] Ibid., pp. 138–39, and Wittgenstein, *Remarks*, p. 155.

sis according to which the sociology of knowledge is unable to account for developments in logic and mathematics.[22]

For Bloor, the elaboration of a sociology of logic is in no respect incompatible with the ability to retain a notion of objectivity that is characteristic of logic and mathematics and dear to Gottlob Frege.[23] The notion of objectivity simply has to be reformulated in Durkheimian terms, according to Bloor, even if this means turning Frege upside down. For Bloor, the objectivity of a concept is nothing but its institutionalization: every concept is inscribed within the framework of a worldview that has been received and transmitted, underwritten and supported, by authorities, and every concept in turn serves to shore up these authorities. In this sense, objectivity is inextricable from sociality—for example, the objectivity of numbers lies in the very institution of the practice of counting. What is at stake for Bloor is the rehabilitation of John Stuart Mill's empiricist conception of the status of mathematics[24]—a conception that had been harshly criticized by Frege[25]—in order to reread the objectivity of symbolic manipulations of mathematics and logic as the institutionalization of physical operations, which would themselves be the authentic underpinnings of mathematical thought.[26]

Bloor's project raises several questions at the outset. First of all, once the equivalence between logical necessity and morality is presupposed, on the basis of what resources can either one be accounted for? More specifically, can a Durkheimian sociology supply the resources to account for potentially competing practices, and, in particular, multiple forms of objectivity among specific groups of logicians? Is the nature of sociality as invoked by Bloor pertinent to the attempt to account for such

---

[22] See David Bloor, "Wittgenstein and Mannheim on the Sociology of Mathematics," *Studies in the History and Philosophy of Science* 4 (1973): 173–91. See also Karl Mannheim, *Essays on the Sociology of Knowledge* (London: Routledge, 1952).

[23] See Gottlob Frege, *The Foundations of Arithmetic: A Logico-mathematical Enquiry into the Concept of Number*, trans. J. L. Austin, 2nd ed. (New York: Philosophical Library, 1953 [1903]).

[24] John Stuart Mill, *A System of Logic, Ratiocinative and Inductive: Being a Connected View of the Principles of Evidence and the Methods of Scientific Investigation* (London: Longmans, 1961 [1843]).

[25] See also the critiques developed in Husserl, *Logical Investigations*.

[26] The question of the naturalist foundation of mathematics is the object of a vast literature in the realm of the cognitive sciences in education. I cite only a few recent works: Uri Wilensky, "Abstract Meditations on the Concrete and Concrete Implications for Mathematics Education," Massachusetts Institute of Technology, Epistemology and Learning Group, Memo no. 12, May 1991; Jean Lave, James Greeno, Alan Schoenfeld, Steven Smith, and Michael Butler, "Learning Mathematical Problem Solving," Institute for Research on Learning, Palo Alto, Calif., Report no. IRL88–0006, August 1988; Idit Harel and Seymour Papert, "Software Design as a Learning Environment," *Interactive Learning Environments* 1 (1990): 1–32.

an object of investigation?[27] In addition, rather than seeking a naturalist foundation for symbolic operations, would it not be more fruitful to look at how social actors associate material operations with symbolic operations, and to investigate the specific effects and stakes of establishing such a relationship?

As we have seen, Bloor undertook several case studies in order to support and develop his analyses.[28] An analysis of Bloor's narratives will allow me to elaborate upon the questions I have just formulated and to assess the scope and limits of Bloor's project in relation to the one developed in this book.

As far as logic is concerned, Bloor's work consisted essentially in stressing its absence of unity, by insisting on the way arguments that have been deemed logical diverge from one society to another and from one period to another. Werner Stark had sought to show the limits of relativism by asserting that ".so far as purely formal propositions are concerned, there simply is no problem of relativity." Taking as an example the assertion that "the whole is greater than the part," Stark maintains that "there can be no society in which this sentence would not hold good."[29] Against this argument, Bloor draws on secondary historical sources to show that the statement in question, the correlate of a logical syllogism, has not always had the same status among mathematicians.[30] According to Bloor,[31] the possibility of putting the set of whole numbers into correspondence with one of its parts (even numbers) served first to prove that the notion of an infinite set was contradictory, then to deny the existence of infinite sets (Augustin Cauchy's thesis), and finally to define infinite sets (for Richard Dedekind, "a system S is said to be *infinite* when it is similar to a proper part of itself").[32]

The way Bloor proposes to address the question of the contradictory character of the sorcerer's status among the Azande people, starting from

---

[27] See the critique developed on this topic in Bruno Latour and Steve Woolgar, *La vie de laboratoire: La production des faits scientifiques*, trans. Michel Biezunski (Paris: La Découverte, 1988), pp. 20–22. [Translator's note: In a foreword to this translation of *Laboratory Life: The Social Construction of Scientific Facts* (Beverly Hills: Sage, 1979), Bruno Latour acknowledges that he himself translated the first chapter "very freely"; the questions raised here do not appear in the original English text.]

[28] These studies are essentially based on secondary historical sources, narratives proposed by other historians of science.

[29] Werner Stark, *The Sociology of Knowledge: An Essay in Aid of a Deeper Understanding of the History of Ideas* (London: Routledge, 1958), p. 163.

[30] Carl B. Boyer, *The History of the Calculus and Its Conceptual Development* (New York: Dover, 1959), p. 296.

[31] Bloor, *Knowledge and Social Imagery*, pp. 135–38.

[32] Richard Dedekind, *Essays on the Theory of Numbers* (New York: Dover, 1963 [1901]), p. 63.

Edward Evans-Pritchard's work,[33] offers a second illustration of his approach to logic. According to Bloor, if Azande logic appears so different from Western logic, it is because institutional frameworks for thought differ from one society to another.[34]

In fact, for Bloor, formal rules and propositions do not constitute real constraints on the production of arguments, the collective management of behavior, or the workings of institutions, inasmuch as the bond between a rule and any case supposed to be governed by that rule always has to be created. For Bloor, the application of a rule or a formal argument in a particular case that has the potential to call into question the stability of institutions to which actors are attached may easily be contested by the production of informal arguments adapted to the situation. In particular, given the stability of the institution of sorcery among the Azande, the anthropologist's invocation of formal contradictions in the definition of the sorcerer's status carries little weight: such contradictions are easily gotten around by "primitives" who bring adequate informal arguments to bear. The study of this case thus allows Bloor to illustrate the way in which an argument may appear quite logical or quite illogical depending on the society in which it is formulated and the institutional constraints that bear on it.

At first glance, these two cases seem to illustrate the existence of cultural and historical factors accounting for variations within logic.[35] But to what extent can this be asserted? The first case constitutes a negotiation between scientists over the status of a logical statement only in that Bloor *himself* attributes a syllogistic translation to the object of the mathematicians he had encountered. In the second case, the author *himself* (like other analysts before him) introduces objects that he attributes to "logic." He *himself* establishes a relationship between these objects and the behaviors and the forms of reasoning that he attributes to the actors.

Thus neither of these two cases retraces a dynamic of production of logical knowledge certified by scientists. Perhaps the historiography of logic did not supply Bloor with materials he could use to carry out such a task?[36] Whatever the case may be, these examples ought to incite us to

[33] Evans-Pritchard, *Witchcraft, Oracles, and Magic.*

[34] Bloor, *Knowledge and Social Imagery*, pp. 138–46.

[35] See also Marcel Granet, *Études sociologiques sur la Chine* (Paris: Presses Universitaires de France, 1953), pp. 95–155. Granet examines the cultural factors contributing to variations in logic and distinguishes a Western mode of logical thought connected to the use of syllogisms from a mode of reasoning by analogy used in China. On this topic, see also François Jullien, *The Propensity of Things: Toward a History of Efficacity in China*, trans. Janet Lloyd (New York: Zone Books, 1995 [1992]).

[36] On this topic, see Imre Lakatos, *Proofs and Refutations: The Logic of Mathematical Discovery*, ed. John Worrall and Elie Zahar (Cambridge: Cambridge University Press, 1976), p. 82.

exercise great vigilance and to be aware that we ourselves risk introducing translations that may lead us to lump together objects that the actors view as distinct, or to attribute to the actors objects that are actually foreign to them. Conversely, we must also take as an object of analysis the translations and relationships brought about by actors, who, like Bloor, may lend themselves to such an exercise in view of specific objectives that must be determined.

Moreover, if in the first case described above Bloor presents transformations of the status of the statement "the part is greater than the whole" without providing causal explanations of a sociological nature, a look at two other cases he borrows from the history of mathematics will give us a better grasp of the modalities of the "institutional" sociology he sought to develop. The first case involves the demonstration of Leonhard Euler's theorem on polyhedra. Bloor starts from Imre Lakatos's description of the way the controversy over Euler's theorem unfolded: the definition of a polyhedron, the statement of the theorem, and its proof were all reformulated several times.[37] Bloor takes up several of the historian's theses on his own account and develops them further.[38] The particular questions they raise are worth examining here.

Seeking to extend the scope of Karl Popper's theses on falsification in mathematics and logic,[39] Lakatos attempts to show, on the basis of a case study focused on Euler's theorem, that mathematics proceeds from a succession of conjectures and refutations. For Lakatos, formal deduction is neither the source of the concrete production of the vast majority of mathematical statements nor a source of certainty. For Lakatos, the idea according to which recourse to formal deduction can be a source of infallibility can only stem from a momentary illusion, one that is historically inscribed in a long series of successive reforms of the canons of rigor in mathematics.[40] Lakatos seeks to show that the formal presentation of a result comes only at the end of a process through which this result is stabilized; the formal presentation itself is nothing more than the registration of past negotiations. These negotiations may consist in proposing

---

[37] Ibid.

[38] See Bloor, *Knowledge and Social Imagery*; Bloor, "Polyhedra and the Abominations of Leviticus," *British Journal for the History of Science* 2, no. 39 (1978): 245–72, or Bloor, "Polyhedra and the Abomination of Leviticus: Cognitive Styles in Mathematics," in *Essays in the Sociology of Perception*, ed. Mary Douglas (London: Routledge and Kegan Paul, 1982), pp. 191–218.

[39] Karl R. Popper, *The Logic of Scientific Discovery* (London: Routledge, 2002 [1934]).

[40] See Lakatos, *Proofs and Refutations*, pp. 55–56, 124–26, and also p. 88, for a denunciation of the profoundly ahistorical character of the work of certain historians of mathematics who reread earlier works in the light of contemporaneous practices that they wrongly judge definitively rigorous.

or rejecting modifications of the result and of the concepts at stake, in transforming a definition to stand up to a counterexample, or it may consist, on the contrary, in retaining the definition while denouncing the "monstrous" character of the counterexample. According to Lakatos, informal thinking always circumvents formal thinking in order to modify the demonstrations, the results, and the meaning of the provisionally stabilized concepts.

After the rise of an intuitionist philosophy of logic and mathematics[41] that revalorized the informal as opposed to the formal (but also the "mental foundations" of logic that had been so roundly denounced by Husserl and Frege),[42] and following upon the research into mathematical heuristics undertaken by George Polya and other mathematicians,[43] Lakatos attempts to take the opposite position from a formalist philosophy of mathematics and logic, which he compares to the study of the dissection of cadavers in the field of human biology, in order to develop a history and a philosophy of mathematical and logical *research* that would be comparable to the study of physiology in the same field of human biology.[44] This approach is of interest in that it restores a historical dimension to mathematical objects, to procedures of proof and the ways they are discussed, illuminating the use in research of resources that are not always celebrated at the time (resources evoked in particular by the somewhat all-purpose expression "informal arguments").

However, the *stylization* of this activity in the form of a perpetual exchange of *arguments* is problematic as such. Like the Roman historians who used fictional dialogues among the protagonists to present a historical narrative, Lakatos in effect re-creates the controversy around Euler's theorem by inventing a dialogue among students gathered in a classroom, each student playing the role of one of the mathematicians involved in the debate. While he does not affirm explicitly that the practices of mathematical research and the actions undertaken in the course of the controversy can be assimilated to such a specific situation, the question that arises is to what extent they are distinct from it. In other words, we may wonder what the nature of this activity is, and in what respects the recounting of such debates may not be reducible to a study of argumentation.

Even though he situates himself explicitly in the research axis established by Lakatos, Bloor for his part seeks to "destylize" the process of negotiating Euler's theorem. In order to account for the unfolding of the

[41] See L.E.J. Brouwer, "Historical Background, Principles and Methods of Intuitionism," *South African Journal of Science* 49 (1952): 139–46.

[42] See Frege, *The Foundations of Arithmetic*, and Husserl, *Logical Investigations*.

[43] See George Polya, *Mathematical Discovery: On Understanding, Learning, and Teaching Problem Solving*, 2 vols. (New York: Wiley, 1972 [1962]).

[44] See Lakatos, *Proofs and Refutations*, pp. 3–4.

debates, the positions adopted by the participants, and, finally, the way that more or less stable "truths" were recognized, Bloor invokes causes of a sociological nature. In "Polyhedra and the Abominations of Leviticus," postulating a structural identity between visions of the metaphysical order and the social order, Bloor judges that a variety of approaches to managing anomalies with respect to a preexisting order may be adopted in mathematics. In his view, the adoption of a given approach is characteristic of the particular society—understood as an institution—in which the mathematicians involved are working.[45]

Bloor uses Mary Douglas's classifications to establish a link between different types of societies (which he characterizes primarily in terms of their greater or lesser degree of social aggregation) and the modes according to which they deal with anomalies.[46] In this way, he distinguishes a society that favors dialectical methods from one that allows a simple coexistence of contradictory results, another that favors the appearance of monsters, and yet another that allows for adjustments of results while also tolerating exceptions. In particular, according to Bloor, a society that allows dialectical exchanges (especially on the basis of proofs and refutations) is an individualist, pluralist, and pragmatic society favoring competition.

To this causalist model Bloor adds a different "institutional sociology" in order to account for the evolution of the controversy over Euler's theorem. For him, if the debates became dialectical only after 1840, it was because of the transformation in university institutions that was seen in Prussia at the beginning of the nineteenth century, as Roy Steven Turner has shown.[47] The introduction of centralized bureaucratic criteria, imposed on the faculty by university administrations, gave higher priority to making discoveries than to writing manuals and encyclopedias, which had formerly been legitimate objects of research. This approach to institutional sociology (resembling that of certain Mertonians in its insistence on the importance of the social system of rewards and the dynamics of competing publications) allows Bloor, as he sees it, to account to a certain extent for the competitive nature of the debates. However, Bloor himself is aware of some of the obstacles that his approach encounters.

First of all, he recognizes the limited ability of macrostructural determinism to account for fundamentally divergent practices of groups and individuals within a given society. In addition, Bloor has significant difficulty applying his reasoning to the data at his disposal. Thus, for example, the university reforms that took place in Prussia have nothing at all to do

---

[45] See Bloor, "Polyhedra" (1978).

[46] See Mary Douglas, *Natural Symbols* (London: Barrie and Jenkins, 1970).

[47] Roy Steven Turner, "The Growth of Professional Research in Prussia, 1818–1848: Causes and Contexts," *Historical Studies in the Physical Sciences* 3 (1971): 137–82.

with certain protagonists in the debates, who lived elsewhere (for example, the mathematician Seidel, who worked in Munich).

Bloor's analysis of a case involving the negotiation of the nature of algebra, in an article titled "Hamilton and Peacock on the Essence of Algebra," encounters similar obstacles. Here he interprets the development of formalist algebra above all as the conquest of independence on the part of mathematics, and thereby of mathematicians, with respect to religious institutions.[48]

Instead of simply invoking the need for much more thoroughgoing empirical research on such cases, we may wonder whether an approach based on different working hypotheses might not help surmount the difficulties Bloor encountered. In the face of a controversy like the one that unfolded around Euler's theorem, should one not abandon such causalism and try to understand how debates are apt, on the contrary, to reconfigure both the mathematical objects and the social organization—in particular, the coalition of actors—involved? In other words, should one not try instead to understand what new order, at once objective and social, more or less lasting and extensive, may arise from a stabilization of the debates?

Moreover, rather than hypothesizing that mathematical practices are wholly determined by a given institutional context, one might wonder what implications the act of bringing together actors endowed with diverse competencies may have for each of them at the very moment of controversy, what adjustments and redefinitions of mathematical practices are called for in the course of their interactions. Finally, one might attempt to study what a potential specificity of mathematical objects allows and disallows in the production of statements and the evolution of debates.

In any event, it was in addressing this latter question that Andrew Pickering and Adam Stephanides were able to propose an alternative description of the emergence of William Rowan Hamilton's algebra. The sociological approach to mathematics that derives from their description warrants close attention at this stage.

Can One Grasp the Role of Networks of Actors and Practices in the Production of a Theorem?

Through their research into the development of Hamiltonian algebra, Pickering and Stephanides illustrate a project in the sociology of mathematics that is very different from Bloor's, although still inspired by Witt-

[48] See David Bloor, "Hamilton and Peacock on the Essence of Algebra," in *Social History of Nineteenth Century Mathematics*, ed. Herbert Mehrtens, Henk Bos, and Ivo Schneider (Boston: Birkhauser, 1981), pp. 202–32.

gensteinian philosophy.[49] Starting from the study of a letter and an excerpt from a notebook kept by Hamilton at the time he was developing the theory of quaternions, the two authors seek to analyze the conceptual practice put into play in his construction.

According to them, this practice, linked to the manipulation of symbolic inscriptions, is in part "objectively" constrained. Nevertheless, it leaves mathematicians with room to maneuver owing to a "social game" in which they participate. To describe this phenomenon, Pickering and Stephanides borrow Ludwig Fleck's terminology,[50] distinguishing between "forced moves" and "free moves." The two authors thus seek to sketch out a sociology of mathematics that responds to two objectives. The first is to do justice to a form of "objective necessity" that characterizes the product of mathematical work. The second is to bring to light conceptual practices that are anchored on the one hand in a series of "cultural and social constraints," and that are determined on the other hand by the individual strategies of the mathematician who exploits the maneuvering room authorized by the mathematical objects themselves.[51]

One of the salient features of this study lies in the fact that its authors succeed in bringing to light the recourse to operations of adjustment that characterize new mathematical objects (at least for the case being studied). However, we may wonder precisely what Pickering and Stephanides mean by objective necessity. Does this notion refer to some resistance of a transcendental, mental, or even material order that would be imposed in the manipulation of inscriptions (as might be suggested in particular by the attention they pay to operations of transcription)? Their article unfortunately does not offer any clear answer to this crucial question.

Moreover, we may wonder whether this approach might not have been called into question, or at least respecified, by an extension of the field of investigation to other historical sources. What narrative might have been produced if the authors had not limited their study to the work of a single scholar on the basis of two documents?

By not considering the mathematical practices of Hamilton's contemporaries or the debates, interactions, and explicit stakes that may have

[49] See Andrew Pickering and Adam Stephanides, "Constructing Quaternions: On the Analysis of Conceptual Practice," in *Science as Practice and Culture*, ed. Andrew Pickering (Chicago: University of Chicago Press, 1992), pp. 139–67. For a further development of this problematics, see Andrew Pickering, *Mangle of Practice: Time, Agency, and Science* (Chicago: University of Chicago Press, 1995).

[50] See Ludwig Fleck, *Genesis and Development of a Scientific Fact*, trans. Fred Bradley and Thaddeus J. Trenn (Chicago: University of Chicago Press, 1979).

[51] Yves Gingras sees this as the reworking of a Piagetian model of resistance and accommodation. See Andrew Pickering, "In the Land of the Blind . . . Thoughts on Gingras," *Social Studies of Science* 29, no. 2 (1999): 307–11; Yves Gingras, "From the Heights of Metaphysics: A Reply to Pickering," *Social Studies of Science* 29, no. 2 (1999): 312–26.

surrounded the mathematician's production, Pickering and Stephanides finally give us a quite specific image of mathematical work: that of an essentially solitary activity. Yet is this image truly accurate? Furthermore, might not devoting attention to other sources have led the authors to take another look at what, in Hamilton's undertakings, must be associated with "objective constraints" or with the mathematician's inscription in a "social game?" And, more fundamentally, had the investigation been carried to these lengths, might it have led to questioning the very possibility of elaborating a categorization of this type?

One of the strengths of Pickering and Stephanides' article is that it helps formulate these questions—questions that have to be raised in the context of trying to account for the progressive development of Elkan's theorem. But since a primary concern of this book lies in studying modes of certification of knowledge in logic, we must also ask whether it is possible to describe the processes by which such a production can acquire "recognition" that is more or less broad-based and lasting (a term to which we shall have to assign a precise meaning); this problem is not really addressed in Pickering and Stephanides' study, whose goal lies elsewhere. In contrast, the question is echoed in Andrew Warwick's analysis of the rise of the theory of relativity, when Warwick focuses on the coordination that took place at Cambridge around the emergence of the theory and the corresponding mathematical constructions. Thus it seems quite pertinent here to spell out the contribution that Warwick's approach may make to the accomplishment of the project developed in this book.

What Role Does the Scale of Adoption of Specific Practices of Demonstration Play in the Dynamics of Recognition of a Result?

Warwick's work sketches out another form of the sociology—and, correlatively, the sociocultural history—of mathematics.[52] For this historian, who draws primarily on the writings of Wittgenstein,[53] Bloor,[54] and the

---

[52] See Andrew Warwick, "Cambridge Mathematics and Cavendish Physics: Cunningham, Campbell and Einstein's Relativity Theory 1905–1911," part 1, "The Uses of Theory," *Studies in History and Philosophy of Science* 23, no. 4 (1992): 625–56, and part 2, "Comparing Traditions in Cambridge Physics," *Studies in History and Philosophy of Science* 24, no. 1 (1993): 1–25. For similar analyses devoted to the history of calculating machines in England at the end of the nineteenth century, see Warwick, "The Laboratory of Theory or What's Exact about the Exact Sciences?" in *The Values of Precision*, ed. M. Norton Wise (Princeton: Princeton University Press, 1995), pp. 311–51.

[53] Wittgenstein, *Remarks*.

[54] See especially the notion of "theoretical practice" developed in David Bloor, *Wittgenstein: A Social Theory of Knowledge* (London: Macmillan, 1983).

sociologist of science Harry Collins,[55] mathematical practices are cultural practices that rely on the learning of skills in order to deal with problems by referring to canonical solutions.[56] The success of a mathematical school then depends essentially on its ability to transmit and impose its tacit local practices on the scale of an international network. The constitution of skills and the processes of transmitting them on the local level, as well as the resources necessary to deploy them on a broader scale, are precisely the objects of the sociohistorical analysis that the author is attempting to produce. In formulating such a project, Warwick considers that he is directly extending work done by Steven Shapin and Simon Schaffer on the history of experimental practices.[57]

Thus the coordination that took place on a broad scale around the theory of relativity is, according to the author, primarily the product, and not the cause, of the emergence of an international community of physicists in the twentieth century. Warwick sees this community as having been able to maintain a consensus around the theory of relativity to the extent that its members had received identical training in the basic techniques of mathematics and physics, wrote and read articles in the same journals, were very mobile and could work together in major international centers, held regular international conferences, and shared their experimental resources.

Starting from this historical representation, Warwick seeks more specifically to account for the coordination that occurred locally in Cam-

---

[55] Warwick is interested in Collins's notion of theoretical practice and in his descriptions of the tacit skills called upon in the context of experimental work. See Harry M. Collins, *Changing Order: Replication and Induction in Scientific Practice* (London: Sage, 1985); see also Collins, *Artificial Experts: Social Knowledge and Intelligent Machines* (Cambridge: MIT Press, 1990). In the latter work, Collins seeks in particular to show that performing arithmetic operations, even on simple calculators, requires a considerable number of tacit skills. For an ethnomethodological approach to these questions based on a study of the way projects in artificial intelligence have been carried out, see Lucy A. Suchman and Randall H. Trigg, "Artificial Intelligence as Craftwork," in *Understanding Practice: Perspectives on Activity and Context*, ed. Seth Chaiklin and Jean Lave (Cambridge: Cambridge University Press, 1993), pp. 144–78.

[56] For complementary analyses of Warwick's work, see Dominique Pestre, "Pour une histoire sociale et culturelle des sciences: Nouvelles définitions, nouveaux objets, nouvelles pratiques," *Annales Histoire, Sciences Sociales* 3 (May–June 1995), pp. 504–6.

[57] See David Gooding, Trevor Pinch, and Simon Schaffer, eds., *The Uses of Experiment: Studies in the Natural Sciences* (Cambridge: Cambridge University Press, 1989); Steven Shapin and Simon Schaffer, *Leviathan and the Air-Pump: Hobbes, Boyle, and the Experimental Life* (Princeton: Princeton University Press, 1985); Barry Barnes and Steven Shapin, *Natural Order: Historical Studies of Scientific Culture* (Beverly Hills: Sage, 1979). Concerning Robert Boyle's view of the place to be granted mathematics in experimental practice, see also Steven Shapin, "Robert Boyle and Mathematics: Reality, Representation and Experimental Practice," *Science in Context* 2, no. 1 (1988): 23–58, and Shapin, *A Social History*

bridge around the emergence of the theory of relativity, and to describe the transformation of the modes of apprehension of a certain number of objects of physics, a transformation brought about by the use of new mathematical methods.[58] To construct his narrative, the author draws on resources located in a transitional space between a heuristic history, in a metalanguage close to that of the actors, and a sociology of learning. He attempts in effect to show how the new heuristics used in the context of the theory of relativity were determined by the competencies acquired by their promoters during their academic training. To this end, Warwick chooses to look very closely into their courses of study, especially within the Cambridge Mathematical Tripos (a demanding university program leading to a certificate of advanced study in mathematics).

The narrative of this historian of physics and mathematics brings an enlightening viewpoint to bear on certain concrete modalities of the work of abstraction.[59] In particular, the author gives us a detailed study of scholarly practices that correspond to the use of graphic methods. While Warwick does not deal in the same way with mathematical proofs that use symbolic inscriptions,[60] we may nevertheless wonder whether a comparable analysis could not be developed in this case.[61]

The examination of the logical demonstrations that we shall encounter shortly will give us an opportunity to go into this question in some depth. In order to do full justice to the specificity of the symbolic manipulations, and to the potentially crucial role of the tiniest details of a proof in the formation of individual and collective representations of the ensuing re-

---

of Truth: Civility and Science in Seventeenth-Century England (Chicago: University of Chicago Press, 1994), pp. 317–22.

[58] On this latter point, see also Peter Galison and David J. Stump, The Disunity of Science: Boundaries, Contexts, and Power (Stanford, Calif.: Stanford University Press, 1996).

[59] Warwick refers in this context to an approach advocated in Bruno Latour, Science in Action: How to Follow Scientists and Engineers through Society (Cambridge: Harvard University Press, 1987). It should be noted that this methodological option is explored further by the author in Warwick, "The Laboratory of Theory." For additional analyses, see also Andrew Warwick, Masters of Theory: Cambridge and the Rise of Mathematical Physics (Chicago: University of Chicago Press, 2003).

[60] In the article in question here, the only demonstration that uses mathematical symbols and that the author considers in any detail is reproduced in an annex with no further analysis. Did Warwick consider that mediations of graphic methods should be treated separately? On this specific point, might we be seeing the influence of Collins's view according to which "the locus of knowledge is not the word or symbol but the community of expert practitioners"? On this point, see Warwick, "Cambridge Mathematics and Cavendish Physics," part 1, p. 633, and Collins, Changing Order, p. 159.

[61] For a problematization of the differentiated use of diagrams and symbols in logic and in mathematics, see Claude Rosental, "De la logique au pathologique," DEA thesis, Paris, Université de Paris I–Sorbonne, 1992, and Brian Rotman, "Thinking Dia-grams: Mathematics, Writing, and Virtual Reality," South Atlantic Quarterly 94, no. 2 (1995): 389–415.

sults, it will be necessary to conduct an analysis of symbolic demonstrations as fine-grained as the one produced by Pickering and Stephanides. In other words, if the quality of the historiographical points made by Warwick[62] depends in large part on his bringing to light practices that have been only faintly visible outside the narrow circles of specialists, it will be necessary, on the terrain of investigation presented in this book, to pursue the task of increasing visibility in other directions, and it will be necessary in particular to consider the practices of symbolic demonstrations as full-fledged objects of empirical and methodological investigation.

Moreover, Warwick's analyses raise important questions as to the modalities and stakes of the constitution and extension of tacit local practices in mathematics. They underline the interest, for anyone studying the dynamics of knowledge production in logic, of looking closely at the nature, history, and geography of the ways in which the work of abstraction is learned, for these factors are apt to determine the approaches taken by the protagonists, at least in part.

Finally, while the author's anchoring of his study in a quasi-disciplinary space is doubtless perfectly legitimate in the case under consideration, it nevertheless raises the more general question of the pertinence of a historical sociology of mathematics that defines the framework for investigation without necessarily problematizing it, within the limits of a field that is first and foremost disciplinary. This question warrants some additional discussion.

Can the Analysis of Demonstrative Practices Be Inscribed Solely within the Framework of the History of a Scientific Discipline?

In an article devoted to the history of analysis at the end of the nineteenth century, Hélène Gispert stresses the interest of a sociological approach to the history of mathematics.[63] Seeking to account for the emergence of a new theory of functions, Gispert studied the links between the subjects of publications by members of the French Mathematical Society and the places where they were trained and where they worked, in an effort to

---

[62] In particular, Warwick succeeds in showing how the expression "theory of relativity" is commonly used to describe research programs and practices that are in fact distinct from one another. I should note that I shall soon have the occasion to show in what way this type of situation can be problematized as such, and apprehended as a consequence of the formation of an authentic "collective statement."

[63] See Hélène Gispert, "Un exemple d'approche sociologique en histoire des mathématiques: L'analyse au XIXe siècle," in *Le relativisme est-il résistible? Regards sur la sociologie des sciences*, ed. Raymond Bourdon and Maurice Clavelin (Paris: Presses Universitaires de France, 1994), pp. 211–20.

characterize the way the influence of various institutions on the directions taken by mathematical research evolved over time.[64] Her analysis is a useful resource inasmuch as it enables her to identify the role of a scientific milieu, the importance of leading academic figures, and—even though the principles of the method she sketches out are not posited in Pierre Bourdieu's terms—phenomena of reproduction in the definition of legitimate fields of research.

Echoing Gispert's work in a programmatic article, Amy Dahan-Dalmedico also urges the development of a social history of mathematics.[65] By this the author means a historicization of the fundamental mathematical concepts and of the heterogeneity of mathematical practices, a historicization based on a study that is not simply limited to the work of an individual or to the history of a concept[66] or theory but also includes an analysis of the "institutional context," viewed as indistinguishable from the "intellectual context." Borrowing Gispert's vocabulary in part, Dahan-Dalmedico argues that this approach needs to be based on the study of a scientific milieu, a community, a set of scientific productions, in order to apprehend a "site," a geography of works, the emergence of a discipline, and the role of leading academic figures. For the author, the social history and the sociology of mathematics are thus not at all synonymous with irrationalism; furthermore, the "harsh dichotomy 'universalism *versus* relativism' " leaves no room to take into account the way rationality is splintered in mathematics.

These reflections could nourish numerous axes of discussion in the context of this book, especially as regards the methodological tools needed to grasp the resources invoked above. However, the choice of the realm of investigation and of its potential restriction to a disciplinary framework is the issue that must be addressed first and foremost. For what cases is such an approach in fact relevant? And, to borrow an expression that the historian of science Norton Wise likes to use to characterize a history immersed in an exclusive focus on a single discipline, in cases where such an approach is not relevant, can one go beyond a "tunnel history" of mathematics?[67]

[64] See also Hélène Gispert, *La France mathématique : La Société mathématique de France (1872–1914)* (Paris: Société française des sciences et des techniques; Société mathématique de France, 1991), pp. 13–180.

[65] Amy Dahan-Dalmedico, "Réponse à Hélène Gispert," in Bourdon and Clavelin, *Le relativisme est-il résistible?* pp. 221–25.

[66] For an example of a work proposing a conceptual history of mathematics, see Amy Dahan-Dalmedico and Jeanne Pfeiffer, *Une histoire des mathématiques: Routes et dédales* (Paris: Seuil, 1986).

[67] Wise, *The Values of Precision.*

By the nature of the response he offers to this last question, Éric Brian clearly highlights at least its fecundity. In *La mesure de l'État*, Brian mobilizes a social and cultural history of mathematics that accounts for the emergence in eighteenth-century France of what will constitute the discipline of statistics a century later, at the crossroads between the rise of the concepts and practices of administrative record-keeping and the rise of mathematical analysis.[68] The sociology that he deploys thus leads him to a detailed study of the practices of analysis in the eighteenth century, especially within the French Royal Academy of Science. Brian shows how this exercise stemmed from practices of decomposition, practices whose origin he traces to an approach to teaching that advocated *exhaustive* methods for *reducing* complex problems to simpler or canonical ones, using procedures of the tree-diagram type to isolate "the things on which pedagogical persuasion" ought to be exercised.[69]

By paying close attention to the material traces of scholarly practice, a process that he sees as opening up a general perspective on the development of a material history of abstraction,[70] Brian succeeds moreover in showing that these practices of decomposition *simultaneously* organized procedures of integration, demonstrative *compositions*, administrative methods for counting people and resources, and charts displaying types of knowledge.[71] Thus he accounts for specific and historically situated mathematical practices and skills, while at the same time making clear the unity of an approach that was spelled out over the course of several exercises and in a variety of spaces.

The social and cultural history of mathematics Brian constructs thus does without recourse to a notion of context or to the description of data whose links to the mathematical practices under consideration remain vague, largely inexplicit, overly indirect, or even simply absent. On the contrary, the author embraces in a single movement what some would separate into "content" and "context," by exploring a set of connections that a "tunnel history" of mathematics would have been unable to grasp.[72]

Bringing to light these connections has a heuristic virtue for (not to say a truly striking effect on) the historical approach, that of making it possible to re-create mathematical practices that are historically situated and difficult to apprehend on the basis of the examination of mathematical

[68] Éric Brian, *La mesure de l'État: Administrateurs et géomètres au XVIIIe siècle* (Paris: Albin Michel, 1994).

[69] See also Brian, "Le livre des sciences,"

[70] Ibid.

[71] Brian, *La mesure de l'État*, pp. 49–93.

[72] See also the analysis of the debate over "internalism" versus "externalism" in Steven Shapin, "Discipline and Bounding: The History and Sociology of Science as Seen through the Externalism-Internalism Debate," *History of Science* 30 (1992): 334–69.

texts alone. In fact, the mathematical texts Brian examined allow for various reading practices; on their own, they do not provide keys to understanding the actors' actual practices—for the historian, these practices are kept at a distance by the existence of very specific contemporary modes of reading. That is why the study of the scholarly use of tables in exercises other than those of analysis ultimately proves fruitful, for example: such an examination permits the reader *in return* to grasp the way in which the work of solving an equation or reducing an integral and the use of *formulas* also stemmed in fact from procedures of the tree-diagram type.[73]

The exploration of such connections has an additional virtue: it extricates us from a rudimentary sociological causalism, by showing that scholars had maneuvering room in which they could configure both their "objects" and the "social context," playing, for example, with classifications that had consequences for the social organization of scientific work in various spaces.[74] In the context of the present book, these observations highlight the importance of studying demonstrative practices by examining their possible declensions and the extent to which these practices correspond to specific types of learning and to researchers' involvement in a variety of exercises. In other words, even while I remain attentive to the potential existence of distinct demonstrative practices conflated under the label "demonstrations," I shall have to ask whether the use of this same term cannot be extended in a pertinent way to a set of practices and actions that are deployed, inconspicuously, in a connected fashion.

At this stage, I need go no farther in examining the works of historians of mathematics who have tried to develop sociological approaches to their object[75] in a more or less problematized way.[76] To prepare for the task of identifying the logical practices and demonstrative mediations

[73] See Brian, *La mesure de l'État*, p. 61: "Beneath the formulas, there are charts."

[74] In this connection Brian speaks of the institutions and their empirical programs as "sedimented classifications" (ibid., p. 111).

[75] The sociological problematic is often (but not always) all the more developed to the extent that the period studied is historically remote.

[76] A large number of relevant works have to be noted here. Not all can be taken as objects of this study, needless to say. Among them, and to cite only a few, see in particular Catherine Goldstein, "Working with Numbers in the Seventeenth and Nineteenth Centuries," in *A History of Scientific Thought: Elements of a History of Science*, ed. Michel Serres (Oxford: Blackwell, 1995), pp. 344–71; Catherine Goldstein, *Un théorème de Fermat et ses lecteurs* (Saint-Denis: Presses Universitaires de Vincennes, 1995); Giovanna Cleonice Cifoletti, "Mathematics and Rhetoric: Pelletier and Gosselin and the Making of the French Algebraic Tradition," Ph.D. diss., Princeton University, 1992; James Ritter, "Les pratiques rationnelles en Égypte et en Mésopotamie aux troisième et deuxième millénaires," Ph.D. diss., University of Paris XIII–Villetaneuse, 1993; Karine Chemla, "Histoire des sciences et matérialité des textes: Proposition d'enquête," *Enquête: Anthropologie, histoire, sociologie* 1 (1995): 167–80; Yves Gingras, "What Did Mathematics Do to Physics?" *History of Science* 39, no. 126 (2001): 383–416; Mehrtens, Bos, and Schneider, *Social History of Mathematics*.

that we shall confront later on, it is time to look closely at some contemporary descriptions of mathematical activity, sociological in inspiration, that were produced by mathematicians looking reflectively at their own practices.

What Demonstrative Resources Are Used for What Recognition?

Of the descriptions of contemporary mathematical practices I have been able to identify that are sociological in inspiration, most have been undertaken by professional mathematicians, not by sociologists,[77] and they have in common their reliance on a sociology that is critical in nature.

In an article titled "Rhetoric and Mathematics," Philip J. Davis and Reuben Hersh contrast the *reality* of research practices in mathematics with a mythical vision that is widespread, according to them, among those who have never carried out research in this area.[78] The two authors focus particularly on those who have known the discipline of mathematics only through their training in high school and college. On the basis of their personal experience, Davis and Hersh denounce a model of rationality that they deem simplistic, according to which the recognition of mathematical demonstrations would be the simple consequence of their intrinsic correctness. What allows a given demonstration to acquire a certain recognition derives, for them, from a complex alchemy.

For these authors, what gives a proof its strength and thus its power to convince depends on many elements: the author's appeal to the reader's

---

[77] See below, however, for my analysis of the works of Donald MacKenzie. Among several other valuable works, Sal Restivo's analyses developing a critical sociology of mathematics should also be noted: see Sal Restivo, *The Social Relations of Physics, Mysticism, and Mathematics* (Dordrecht: D. Riedel, 1983); Restivo; *Mathematics in Society and History: Sociological Inquiries* (Dordrecht: Kluwer, 1992). See also Randall Collins and Sal Restivo, "Robber Barons and Politicians in Mathematics: A Conflict Model of Science," *Canadian Journal of Sociology* 8 (1983): 199–227; Sal Restivo, Jean Paul van Bendegem, and Roland Fischer, *Math Worlds: Philosophical and Social Studies of Mathematics and Mathematics Education* (Albany: State University of New York Press, 1991); Luke Hodgkin, "Mathematics as Ideology and Politics," in *Radical Science Essays*, ed. Les Levidow (London: Free Association Books, 1986), pp. 173–197; Marcia Ascher, *Ethnomathematics: A Multicultural View of Mathematical Ideas* (Pacific Grove, Calif.: Brooks/Cole, 1991); and Helen Verran, *Science and an African Logic* (Chicago: University of Chicago Press, 2001).

[78] Philip J. Davis and Reuben Hersh, "Rhetoric and Mathematics," in *The Rhetoric of the Human Sciences: Language and Argument in Scholarship and Public Affairs*, ed. John S. Nelson, Allan Megill, and Donald N. McCloskey (Madison: University of Wisconsin Press, 1987), pp. 53–68. See also Philip J. Davis and Reuben Hersh, *The Mathematical Experience* (Boston: Birkhauser, 1981); Philip J. Davis and Reuben Hersh, *Descartes' Dream: The World according to Mathematics* (San Diego: Harcourt Brace Jovanovich, 1986).

intuition, the images, the meta-arguments, the lack of counterevidence, the dynamic of reference through the invocation of results that may still be unpublished, the appeal to the reader's trust when the author chooses not to justify arguments characterized as too cumbersome, and the dynamics of interactions. By way of illustration, the two authors evoke the impossibility of expressing doubts in the face of mathematicians who are known authorities, the desire to avoid appearing inappropriate or ridiculous in front of an audience, and the tacit skills and practices involved, whether these are local or shared by members of the same subdiscipline. According to Davis and Hersh, these elements are crucial, for as a general rule mathematical proofs are not justified step by step in formal logic. There are several reasons for this: a logical translation is generally impossible to carry out in practice, and even if it could be accomplished, the fundamental notions of logic are highly problematic, the correctness of the translation would remain to be demonstrated, and in any event no one would be interested in the proof, given its illegible character (which would necessitate a new appeal for trust).

As Davis and Hersh see it, the hope that arose in the twentieth century that mathematical proofs could be formally verified by computer was given up for similar reasons (the difficulty and length or even the impossibility of a translation into formal logic; doubts about the principles and modalities of carrying out translations; lack of interest on the part of mathematicians in yoking themselves to such a task). Under these conditions, mathematical practice has not evolved very much.

Nor does the publication of a proof constitute a simple guarantee of its correctness for our authors: in their view, any mathematical article sent in to a journal can be accepted or rejected according to the reviewers to whom it is submitted; the ability to judge a proof's correctness and to discern its problematic aspects and its points of interest stems from a corporatist knowledge that is often inaccessible to individuals working outside of a given subdiscipline; the reviewers generally carry out partial verifications and trust the author for the rest of the proof; the author's reputation, the representations of his or her degree of reliability and originality, and the evaluators' perception of the research field covered by the journal are all important elements in the acceptance or rejection of an article.

Davis and Hersh argue that, once published, papers are not widely read: their readership is generally limited to the author's own immediate circle and a few dozen researchers in the best of cases. This is the moment at which it becomes interesting for mathematicians to point out an error, to offer an extension, a generalization, or an application of a result, or even to propose other proofs and suggest connections with other theorems. The more or less lasting recognition of a theorem's correctness is the consequence and not the cause of going through all these trials.

Other mathematicians have produced descriptions similar to the one we have just examined.[79] In an article published in 1979, three mathematicians and computer scientists, Richard A. De Millo, Richard J. Lipton, and Alan J. Perlis, stress the essential role played in the elaboration and rewriting of proofs and, correlatively, in the capacity of proofs to arouse interest and be recognized to a greater or lesser degree, by the interactions in which the authors of demonstrations are engaged.[80] According to these scientists, in actual practice theorems are not verified in formal logic: either the operation would be impossible or the proof would be unreadable. In the latter case, only blind trust in the author of a formal verification would make it possible for readers to accept the proferred results.

Given these conditions, for De Millo, Lipton, and Perlis the recognition of the correctness of a theorem in mathematics stems above all from social factors. The three authors assert that such recognition can only be more or less widespread and more or less lasting, and that it can be obtained only gradually, passing through several stages such as the use of the theorem in the work of other mathematicians, translations of the result(s), generalizations, links with other statements, and so on. According to these authors, theorems rarely succeed in passing through these stages, for in practice many are disallowed, thrown into doubt, or simply ignored by other mathematicians.

The formulation of critical analyses of mathematical activity by researchers working in the discipline is actually a very ancient practice, even if it has not always taken the form of descriptions from a sociological perspective. In the twentieth century in particular, developments in logic that have been perceived by some mathematicians as an "introduction to

---

[79] See the thesis of the mathematician Raymond Louis Wilder, close to that of Lakatos on the illusion of certainty and the end of the history of rigor in mathematics: "Obviously we don't possess, and probably will never possess, any standard of proof that is independent of time, the thing to be proved, or the person or school of thought using it. And under these conditions, the sensible thing to do seems to be to admit that there is no such thing, generally, as absolute truth [proof] in mathematics, whatever the public may think" (cited in Morris Kline, *Mathematics: The Loss of Certainty* [New York: Oxford University Press, 1980], p. 314). See also Raymond L. Wilder, *Mathematics as a Cultural System* (Oxford: Pergamon, 1981). With a more developed sociological perspective, see also Charles S. Fisher, "The Death of a Mathematical Theory: A Study in the Sociology of Knowledge," *Archives for History of Exact Sciences* 3 (1966): 137–59; and Charles S. Fisher, "Some Social Characteristics of Mathematicians and Their Work," *American Journal of Sociology* 78 (1973): 1094–1118.

[80] See Richard A. De Millo, Richard J. Lipton, and Alan J. Perlis, "Social Processes and Proofs of Theorems and Programs," *Communications of the Association for Computing Machinery* 22, no. 5 (1979): 271–80.

new canons of rigor" have led others to denounce the pretentions of formal logic. Henri Poincaré belonged to the latter group,[81] and so did Henri Léon Lebesgue, who asserted that "logic gives us reasons to reject certain arguments; it cannot make us believe an argument."[82] What some have seen as a turning point for rigor has led certain mathematicians to question the practice and the nature of mathematical proofs. Thus Godfrey Harold Hardy observed "that there is, strictly, no such thing as mathematical proof; that we can, in the last analysis, do nothing but *point*; that proofs are what Littlewood and I call *gas*, rhetorical flourishes designed to affect psychology, pictures on the board in the lecture, devices to stimulate the imagination of pupils."[83]

Beyond their denunciatory character, declarations such as these open up quite interesting perspectives on the development of a sociology of mathematical and logical practices. Must we prepare ourselves, as Hardy in particular implied, to call into question (after a methodological investigation and not merely a verbal denunciation) the expectation, a highly sedimented one in the philosophy of science, that a dichotomy can be established between showing and demonstrating?

Statements like Hardy's suggest, in any event, that demonstrative practices proceed from the deployment of a set of polymorphic resources, whether these are incorporated into the texts of proofs, mobilized orally, shaped in the course of interactions, or distributed in more or less widespread habits. The mathematicians' descriptions cited above call for detailed analyses of specific cases. The analyst would need to assess the importance of each of the mediations evoked and establish a hierarchy of their respective roles while situating them within dynamics of production of certified knowledge. Testimonies such as those I have cited raise absolutely crucial questions about the processes that lead to the publication and recognition of a result. What tests must a theorem pass in order to win legitimacy? What signifies a "correct," "recognized" result for the actors involved? How do the actors articulate such notions, if they find them relevant at all?

Before dealing with questions like these, it will be useful to look at some scattered studies that focus on logical practices, or at least suggest ways of conducting empirical research on logic in action. The fact is that, while I have been able to identify some texts by mathematicians that deal with

[81] See Lakatos, *Proofs and Refutations*, p. 81.

[82] Henri Léon Lebesgue, *Leçons sur l'intégration et la recherche des fonctions primitives* (Paris: Gauthier-Villars, 1928 [1904]), p. 328.

[83] Godfrey Harold Hardy, "Mathematical Proof," *Mind* 38 (1929): 18. On this subject, see also Jean Paul van Bendegem, "Non-formal Properties of Real Mathematical Proofs," in *Philosophy of Science Association 1988*, ed. Arthur Fine and Jarrett Leplin, vol. 1 (East Lansing, Mich.: Philosophy of Science Association, 1988), pp. 249–54.

the concrete aspects of their activity, I have been unable to locate equivalent descriptions by logicians of their own practices. My search for work by sociologists and historians on the question had not been much more fruitful when I started to write this book. Lakatos may have put his finger on one explanation for this situation when he judged that "unfortunately, even the best historians of logic tend to pay exclusive attention to the *changes in logical theory* without noticing their roots in *changes in logical practice.*"[84]

However this may be, readers who believe that the research process in the experimental sciences and in mathematics constitutes an object of sociological investigation may extend this assumption to the formal and symbolic manipulations at work in the exercise of these sciences. Even so, they will not necessarily suppose that the *production* of knowledge in logic can itself be apprehended from a sociological perspective. The foundations of scientific activity have often been attributed to "logic" in the framework of architectonic visions of knowledge or specifically of "logics of science." Thus some readers might spontaneously judge that a sociogenesis of the activity of the so-called experimental sciences, or even of mathematics, is in no way incompatible with the fact that these activities "rest" ideally on "logic," which for its part cannot be the object of sociological investigations.

The argument according to which a sociology of the sciences could only have meaning to the extent that it would limit itself to the study of certain quite specific disciplines, such as medicine, biology, physics, and perhaps mathematics, with the categorical exclusion of "logic," has to be examined within the context of this study. Thus, even before I present the results of my own research, the recent work I have been able to identify that puts logical practices in the foreground requires careful attention.

QUESTIONS RAISED BY SOME STUDIES THAT FOCUS ON, OR FORMULATE A RESEARCH PROGRAM TO ADDRESS, PRACTICES IN LOGIC

In *Words of Power*, Andrea Nye attempts to produce a social history of logic from classical antiquity to our own day.[85] She rereads this history as a series of enterprises that seek to exclude certain groups or reduce them to silence. For Nye, Aristotle's logic was designed to exclude non-Greeks,

---

[84] Lakatos, *Proofs and Refutations*, p. 81 n. 4.

[85] Andrea Nye, *Words of Power: A Feminist Reading in the History of Logic* (London: Routledge, 1990). For a critical analysis of this book, see John Batali, "The Power and the Story," *Postmodern Culture* 2, no. 2 (1992).

medieval logic to exclude non-Christians, Frege's to exclude non-Aryans. She places particular stress on the fact that Aristotle's syllogisms were intended in the first place for professional dialecticians so they could impose their arguments in court or in political assemblies; she also maintains that Frege developed his logic in conformity with his own antiliberal, anti-Semitic, ultraconservative positions.[86] He expressed these positions in his political notebooks, and his teaching practices followed suit: Frege would not have tolerated the slightest intervention from his students during or after his classes.[87]

One interesting aspect of Nye's work is that she seeks to grasp the development of distinct logical systems as they are implicated in a series of historically situated projects and spaces. However, one may wonder whether the fact that she focuses her study of the history of logic on elements that serve to shore up a political critique does not lead the author to place certain characteristic dimensions of the scholarly practices involved outside the field of analysis. Attention to those dimensions would make it possible, in particular, to grasp the modalities of production of certified knowledge in this area. Conversely, it could be interesting to speculate about the potential role that the denunciations made or the political-cultural projects advanced by actors in logic themselves might play in the context of the dynamics of accreditation of knowledge. The study of the emergence of Elkan's theorem will give us an opportunity to test the degree of pertinence of such questions.

For now, I must also evoke the way in which logical production has been approached from a radically different perspective: that of the ethnomethodology of mathematics, in the book Eric Livingston devoted to the proof of Gödel's incompleteness theorem.[88] While readers of this work (at least those I have been able to identify) commonly but erroneously see in it a paraphrase of Kurt Gödel's work,[89] Livingston's work may be apprehended in another way. I would suggest that the author is seeking to account, through his analysis of Gödel's text, for the resources and demonstrative practices the mathematician brings into play, while emphasizing what a line-by-line study of the proof entails.

[86] Nye, *Words of Power*, p. 152 n. 6.

[87] Ibid., p. 159.

[88] Eric Livingston, *The Ethnomethodological Foundations of Mathematics* (London: Routledge, 1985). See also Livingston, "Cultures of Proving," *Social Studies of Science* 29 (1999): 867–88.

[89] The most radical ethnomethodology seems to recognize no better metalanguage than the terms of the actors' practical technical competence. This would make it possible to understand the nature of Livingston's descriptive vocabulary, which some have seen as paraphrase. On this methodological question, see Latour and Woolgar, *La vie de laboratoire*, p. 26, and Lynch, *Art and Artifact*.

The ethnomethodological program formulated by Harold Garfinkel[90] on the basis of his reading of Alfred Schütz[91] was aimed at describing a plural rationality. When it came to describing scientific activity, among the ways rationality could be deployed Garfinkel chose to retain the one corresponding to the category of commensurability with the principles of "formal logic."[92] In a sense, Livingston questions the unity of this category and at the same time tries to explode and reconstruct the category by associating it with a plurality of practices. This important question will have to be raised again in the framework of the present book, as we shall have to be sensitive to the various declensions of the exercise of logic and of the demonstrative practices considered.

However, as in the case of the study by Pickering and Stephanides, we may wonder whether Livingston's analysis of Gödel's text might not have been profoundly modified if the field of inquiry had been extended to other sources. To what extent would the description of resources and practices have been retained if Livingston had looked at the practices of the logician's contemporaries, and also at the interactions, debates, and explicit stakes that surrounded the drafting of the incompleteness theorem?

While each of the authors I have mentioned did finally make a very specific methodological choice that may well intrigue some readers, it is quite instructive at this stage to speculate a bit about its origins. As we have seen, the idea that one could actually carry out an empirical study of logic while it is being practiced has simply not been envisaged by many philosophers. Some have viewed such an undertaking as of very limited interest, while others have written it off as completely irrelevant and totally detached from their philosophical problematics.

Given the fact that projects for sociological studies of logic have generally been developed *in reaction* to such philosophical representations, certain analysts have *begun* by plunging into discussions of principle bearing on the feasibility of such a study. In many cases, confronted by a substantial philosophical literature, the authors seem to have expended all their energy on this task, to such an extent that they hardly go beyond the stage of discussing principles.

Moreover, the resonance of certain forms of philosophical idealization of logical activity in the representations of the analysts themselves may have constituted a significant stumbling block for the development of em-

[90] Harold Garfinkel, *Studies in Ethnomethodology* (Englewood Cliffs, N.J.: Prentice Hall, 1968).
[91] See Alfred Schütz, "The Problem of Rationality in the Social World," *Economica* 10, no. 38 (1943): 130–49.
[92] For a critique of Garfinkel's reading that emphasizes the unsatisfactory nature of his criteria, see Latour and Woolgar, *Laboratory Life*, p. 153 (*La vie de laboratoire*, p. 149).

pirical investigations of this type of object: it was not a *given* that logical activity could be investigated *in any way other than* as an immaterial process or a solitary enterprise unfolding essentially in scholars' minds. Consequently, the possibility that one could carry out *in situ* observations of the material interactions and practices (writing, among others) that characterized groups of actors in logic was not at all obvious.[93] Under these conditions, the pursuit of research consisting in various forms of *textual study*, whether they belong to a semiotic or ethnomethodological tradition, a possibly Piagetian approach, or the genre of the philosophical essay, constituted a preestablished path for the authors.[94]

The situation was not very different in the cognitive sciences, moreover. As Edwin Hutchins has ably shown, the resonance of certain forms of idealization of logical activity in cognitive science has helped make this same activity an abstract one as seen by researchers.[95] In particular, the reprise of certain models of symbolic calculus (like Alan Turing's) and of forms of essentialism related to work in logic has, according to Hutchins, caused cognitive scientists to lose sight of the concrete dimension of this activity.[96] For Hutchins, replacing the mind by a computer in cognitive science has had as a correlative the disappearance of the hands and eyes of the individuals manipulating symbols. According to him, this activity has been placed solely in the "heads" of human beings, all the more easily in that the cognitive sciences have tried more generally to "load" the human mind with a mass of faculties specific to it; this approach has been developed in reaction to behaviorism's representations of a human being

[93] The pursuit of a sustained reflection on the notion of self-evidence in logic and my own insertion into a research environment that was exceptional on more than one count are among the factors that have given my methodological approach its specificity.

[94] The first reaction of one specialist in the semiotics of mathematics, when I met him and told him that I had undertaken an empirical study of the work of logicians, illustrates this phenomenon quite well. He reacted with astonishment, even though he himself had always emphasized the material dimension of the graphic activity of mathematicians. Neither more nor less than most of the other interlocutors I encountered during my investigation, this author *did not realize* that one could observe the work of logicians *in situ*, and he remained quite intrigued by my project.

[95] See Edwin Hutchins, *Cognition in the Wild* (Cambridge: MIT Press, 1995). In this work, the author looks at modes of piloting a big American battleship. He studies the way certain formalisms (although different from logical formalisms) can be mobilized for the purpose of coordinating a group of actors. For a critical account of this work, see Bruno Latour, "Cogito Ergo Sumus! Or Psychology Swept Inside Out by the Fresh Air of the Upper Deck . . . ," *Mind, Culture, and Activity* 3, no. 1 (1996): 54–68.

[96] Let us note, however, that these descriptions scarcely apply to cultural or contextual approaches to cognition. See especially Michael Cole and Sylvia Scribner, *Culture and Thought: A Psychological Introduction* (New York: Wiley and Sons, 1974). For an example of the "contextualization" of mathematics, see Jean Lave, *Cognition in Practice: Mind, Mathematics and Culture in Everyday Life* (Cambridge: Cambridge University Press, 1988).

rich in behaviors even though endowed with a brain that may in the extreme case be empty.[97]

However, if it was difficult to imagine observing logic in the making, and especially the "standard" production of theorems, this was not the case for certain specific mediations of logical activity. In computer science, for example, logic is used in ways that can be much more easily perceived in their material dimension, and thus can be seen as suitable objects for empirical study. Thus it is hardly surprising that studies finally appeared on this terrain almost simultaneously. Some of these studies were carried out by Donald MacKenzie, among others, some by myself.[98]

In an article published in *Social Studies of Science*, Donald MacKenzie juxtaposes two very focused studies dealing with the development of microprocessors.[99] Drawing on some interviews and the analysis of a certain number of documents, MacKenzie foregrounds debates proper to the realm of computer science, debates that put mathematical and logical knowledge into play.[100] He first shows that the standard for floating-point arithmetic adopted by the American Institute of Electrical and Electronics Engineers in 1985 was not simply deduced from so-called classical arithmetic, but negotiated over a long period of time in relation to the strategic interests of the major builders and their spokespersons. In the same article, MacKenzie describes the unfolding of a controversy about the reliability of a microprocessor. The manufacturer of a microprocessor had presented its architecture as having been subjected to formal verification, and client companies had challenged this claim, thereby launching a debate over the notion of a formally correct proof.[101]

In support of these two cases of development and implementation of computer systems, MacKenzie tries to show that mathematics can be the

---

[97] See especially Hutchins, *Cognition in the Wild*, p. 361, and Daniel C. Dennett, *Consciousness Explained* (Boston: Little, Brown, 1991).

[98] See in particular Claude Rosental, "The Art of Monstration and Targeted Doubt: The Anthropology of Mathematical Demonstration," working paper, Centre de Sociologie de l'Innovation, École Nationale Supérieure des Mines de Paris, 1993.

[99] See MacKenzie, "Negotiating Arithmetic, Constructing Proof."

[100] Other descriptions of these practices can be found in the following works devoted to the history of computer science, among others: Mary Croarken, *Early Scientific Computing in Britain* (Oxford: Clarendon Press, 1990); Herman Heine Goldstine, *The Computer from Pascal to von Neumann* (Princeton: Princeton University Press, 1972); Andrew Hodges, *Alan Turing: The Enigma* (New York: Simon and Schuster, 1983); William Aspray, *John von Neumann and the Origins of Modern Computing* (Cambridge: MIT Press, 1990). For a synthetic article on the history of computer science and calculus, in relation to that of logic and the cognitive sciences, see Daniel Andler, "Calcul et représentation: Les sources," in Andler, *Introduction aux sciences cognitives*, pp. 9–46.

[101] See also Donald MacKenzie, "The Fangs of the VIPER," *Nature* 352 (Aug. 8, 1991): 467–68.

object of debates and negotiations. Although the author does not spell out what he means by "sociology" or "sociology of mathematics," this study is clearly compatible with a Bloorian approach to mathematics, one that exposes variations and divergences in mathematical theories, notions, and statements, stressing the way the latter are *taken for granted* locally and temporarily. After focusing on debates in the realm of statistics,[102] MacKenzie clearly perceived the development of computer science as a new privileged object for analyzing mathematics viewed as more or less ephemeral social constructions.[103]

MacKenzie's work raises interesting questions for the development of a sociology of mathematics and logic, even if the problematics that underlies his studies is perceptibly out of alignment with the one developed in this book. At this stage, it would not be useful to go back over the question of social relativism or speculate about the interest of attempting to pin down, in the framework of the present study, what the potential specificity of logical objects allows and disallows in the production of new statements.

In contrast, it is tempting to wonder, in the wake of MacKenzie's studies, what repercussions the rise of computer systems and their uses may have on demonstrative practices in logic. One would have to try to grasp not only how logic can be "negotiated" in the course of developing such systems, but also what specific demonstrative procedures can be used to exhibit the properties of a logic in these systems, and what possible new demonstrative registers can be put to work owing to the existence of such systems.

Moreover, it seems worthwhile to try to determine to what extent empirical investigations may be also conducted into what is commonly perceived as the "heart" of logical activity, namely, the production of theorems in symbolic languages. And it appears fruitful to inquire whether "empirical investigations" on this topic have to be limited to textual analysis and interviews, or whether such investigations may also take the shape of ethnographical observations.

To conclude this brief examination of works that, despite their somewhat heterogeneous character, are of direct interest for the problematics of this book, it will be useful to look closely at the program Bruno Latour proposes for the empirical study of logic in action. I have referred to the fact that, for Latour, the processes of abstraction constitute concrete phenomena that can be grasped in their materiality.[104] In *Science in Action*,

---

[102] Donald MacKenzie, *Statistics in Britain, 1865–1930: The Social Construction of Scientific Knowledge* (Edinburgh: Edinburgh University Press, 1981).

[103] For a later publication on the same topic, see in particular Donald MacKenzie, *Mechanizing Proof: Computing, Risk, and Trust* (Cambridge: MIT Press, 2001).

[104] Latour, "Sur la pratique des théoriciens."

Latour defines a set of objectives for an anthropological study of what he calls formalism.[105] Latour uses this term to refer to the set of representations of the world, or rather of its transformations in the form of inscriptions (graphic, formulaic, printed, and so on). According to him, inscriptions have a twofold interest for those who produce them: they make it possible both to retain "enough things" from the elements taken as objects so as to be able to act on them from a distance by bringing together pertinent representatives of these things in a single place, and to keep "as few things" as possible in the grouping so as to be able to bring together only the representatives without having to mobilize the things themselves.[106]

The author uses the term "formalism" to designate mathematics and logic *in particular*. According to him, mathematical or logical formalism is precisely what can bring together in a single place (called a center of calculation), potentially reducible to a graphic representation on a piece of paper, transformations of elements on which control can be exercised from a distance, since the effects of the smallest action carried out on inscriptions are magnified on the elements represented.[107]

The fact of *transforming* disconnected elements (such as "the concerns of households," the "behavior of amino acids and wooden spheres rolling on inclined planes") into a common *matter*, that of material inscriptions (formalism is therefore material), in order to subject them to a homogeneous treatment—mathematical, for example—constitutes a source of novelty for Latour. Indeed, this approach authorizes both the establishment of unprecedented relationships and a displacement of the centers of action. For Latour, the recourse to formalism thus represents a source of innovation, entailing a displacement of centers of action, mobility, and new connections; it participates in the construction and reconfiguration of the networks present in the world, and it even represents one of the most effective modes for creating associations. In this context, logic, sometimes identified by the author by means of the expression "it is logical," constitutes the very prototype of the (rare) success of an attempt to establish an association among elements through the mobilization of a formalism.[108]

According to Latour, in order to construct an anthropology of formalism, it is thus unnecessary to create artificial links between formalism and

---

[105] The meaning Latour gives this expression has to be distinguished from most of those it has been given in the history and the philosophy of logic. In particular, this term sometimes refers to the approach to logic developed by the school known as formalist, especially by Hilbert (I shall return to this point).

[106] Latour, *Science in Action*, p. 243.

[107] Ibid., pp. 244–45.

[108] Ibid., pp. 57–58, 205–6.

material operations or society; it suffices to follow the associations that formalism itself constitutes.[109] To this end, Latour advocates the empirical study of the processes of *formalization* (*mise en forme*) or abstraction (he uses the terms synonymously): this is perfectly possible within the framework of his model, since both the elements represented and the formalism itself are material in nature.[110]

We thus come back to his project for the material study of abstraction, which can be carried out, he asserts, by observing the working of centers of calculation, then by circulating among the various sites connected to those centers through operations of formalization.[111] Moreover, as he himself has followed the constitution of cascading representations in the experimental sciences, comparing them metaphorically to the productions of a refinery, Latour considers that he has already succeeded in studying processes of abstraction.[112]

This program for research on formalism offers several elements for reflection in relation to the development of the problematics of the present book.[113] It emphasizes the value of taking a material approach to the writing done by researchers and of carrying out on-site observations of their activity. It incites observers to be sensitive to all the connections made or mobilized during the processes of graphic production, and to circumscribe the novelty that results. Moreover, it raises the question of the place to be given to the sequence of operations of formalization for the study of logic as it is being produced.

It is important to stress the fact that one of the salient features of *science in action* consists in reinterrogating a presentation of science that places logic at the origin of scientific activity as a whole. By doing the opposite, by studying the question of formalism (in Latour's sense) last and no

[109] Ibid., pp. 245–46.

[110] From this viewpoint, Éric Brian's project of developing a material history of abstraction may itself be connected in a formal way with that of Bruno Latour. See Brian, "Le livre des sciences," p. 94. Similarly, see the analysis of the role of tables in economics by Jean-Claude Perrot, *Une histoire intellectuelle de l'économie politique (XVII–XVIIIe siècles)* (Paris: École des Hautes Études en Sciences Sociales, 1992). See also the work of "rematerialization" of geometry carried out by Michel Serres in "Gnomon: The Beginnings of Geometry in Greece," in Serres, *History of Scientific Thought*, pp. 73–123. Finally, see also the propositions for rewriting the history of logic formulated in Hayward R. Alker, "Standard Logic versus Dialectical Logic: Which Is Better for Scientific Political Discourse?" Twelfth World Congress of the International Political Science Association, Rio de Janeiro, August 8–15, 1982, p. 29.

[111] See Latour, *Science in Action*, p. 243.

[112] Ibid., p. 241.

[113] In comparison, see also the research program bearing on the development of a sociology of formalism that is set forth in John Bowers, "The Politics of Formalism," in *Contexts of Computer Mediated Communication*, ed. Martin Lea (Hassocks: Harvester, 1992), pp. 232–61.

longer seeking to mobilize any notion of the foundation of science, Latour is attempting to show that this question constitutes just one of the numerous elements that must be apprehended to account for scientific activity. In so doing, he is also calling back into question the idea that the work of logicians grows out of *exceptional* and *ultimate* practices that cannot be observed in other exercises, whether they are considered as scientific or nonscientific. This position is in conformity with his notion of formalism, which grants no particular privilege a priori to the production of logical inscriptions in relation to the production of other inscriptions, and that thus constitutes a tool for decentering the gaze that analysts may be able to use.

The question that arises here, then, is the following: to what extent does what may appear central in contemporary uses of formalisms for the production of biological facts, statistical assertions, or meteorological predictions—that is, operations of formalization—remain central when one studies the production of certified knowledge in logic? Does the contemporary activity of actors in logic consist above all in formalizing, or does it also proceed from other dimensions, other practices, that may be quite disparate?

As I have already noted, an essential dimension of the work done by the researchers in logic I have observed lay in the work of demonstration itself; in what follows, I shall attempt to give this assertion as precise and fully developed a meaning as I can. But while questioning the possible centrality of the work of formalization in logical activity, I shall also need to ask whether we should privilege a model that places logical formalism in central nodes of networks, or whether we should instead envisage a less contrasted situation: if specialists in logic can control numerous elements from a distance, do they not themselves depend for their activity on other centers (not necessarily centers of calculation)?

We shall have to look into this question while taking quite particularly into account the possibility of a situation of competition between logical formalisms, correlating with competition between practitioners in the field, that may lead to a more complex arrangement of formalisms than the one that would result from simple cascades. In this case, we shall have to ask whether the capacity of a logical formalism to succeed in bringing elements into relation in everyone's eyes is not severely limited by the need to acquire a high level of competence in order to apprehend the formalism. Finally, then, we shall have to ask to what extent the formalist activity studied in this book (which may not stem solely from an activity of formalization) lends itself to being characterized not simply in terms of coordination, but also and especially, as we have evoked it, in terms of demonstration.

Conclusion

The set of questions I identified earlier gains from being raised on the basis of my own empirical investigations. We shall not need to review all those questions in detail, of course, but it may nevertheless be useful to reformulate them in their broad outlines.

The goal of my research is not to try to determine to what extent "logic" is used as a "mode of reasoning" in everyone's daily life. I shall seek rather to describe a specific process of production of certified knowledge in logic. I shall not give a restrictive a priori definition of what "logic" entails only to confront it later with empirical definitions. On the contrary, it will be interesting to consider in a first phase whether "logic" as presented by certain philosophers, ethnologists, or psychologists is the same as the logic set forth by the actors we shall meet. We shall consider whether, for this second group, logic rests, for example, on the ideal necessity of logical principles such as the principle of contradiction, or whether it is the object of distinct conceptions and practices. It may then be fruitful to seek to understand in what respect several different philosophies of logic correspond in fact to diverse modes of the exercise of logic among practitioners.

It will be important, in addition, to remain open to a possible plurality of actors in logic—I shall not reduce these at the outset to a single "knowing subject"—as well as to the potential diversity of its objects. Moreover, to describe the process of production of certified knowledge in logic under study in this book, it will be important to take care to avoid accounting for symbolic manipulations by associating them artificially with material operations or characteristics of society while asserting that the former are only the "reflection" of the latter. On the contrary, we shall have to pay attention to the specific associations mobilized or realized in the course of the process, while distinguishing the establishment of equivalences performed by the actors from those that may be carried out by the analyst.

Furthermore, it will be necessary to pin down precisely the respect in which debates linked to the emergence of a logical theorem cannot be reduced to a mere exchange of oral arguments. To this end, a special effort will be made to apprehend the material dimension of the protagonists' work. In particular, it will be important to analyze their writing activities, the potentially antagonistic skills mobilized for the occasion, the role of interactions and the declared stakes involved in the various undertakings of the participants, whatever their nature.

In addition, we shall have to examine the degree to which the more or less extensive character of the practices brought into play, as well as the smallest details of the demonstrations, may determine the evolution of the process being studied. Demonstrative practices will be interrogated in relation to the potentially multiple exercises in which the actors are involved, but also in relation to the varied resources and registers used by the researchers to support their own points of view—whether these elements are presented in the texts of proofs, brought into various types of oral and written interventions, or shaped during interactions. On this occasion, it will be crucial to ask to what extent a dichotomy generally operated between showing (*montrer*) and demonstrating (*démontrer*) is pertinent, and to what extent the formalist activity studied can be identified with an activity of putting into form, or formalization. Finally, we shall have to consider the question of the meaning that can be attributed to the notion of "recognition" of a logical theorem, by analyzing the various tests that may be reserved for the latter so it can achieve the status (still to be specified) of certified knowledge.

Chapter 2

SPACES AND TOOLS FOR EXCHANGE

I N EARLY 1994, as noted above, a researcher in the field brought to my attention an electronic forum called "comp.ai.fuzzy," a Usenet newsgroup devoted to the theme of fuzzy logic. During the previous months, the forum had been the site of debates over Charles Elkan's article. As we shall see shortly, these exchanges are highly relevant to the dynamics under study here; the comp.ai.fuzzy newsgroup thus serves as a useful entry point and an initial "observatory" for viewing logical practices at work.

After analyzing in part 2 the controversy stirred up in the newsgroup by the publication of Elkan's theorem, I shall attempt to situate the role of that controversy more precisely in part 3. To this end, I shall examine the protagonists' parallel actions and then return to the larger context, looking at the phases that preceded and followed the period of debates. For now, in order to offer a concrete grasp of the way the exchanges took place in the electronic forum, I need to provide some details about the forum's procedures and uses.

PRELIMINARY INFORMATION ABOUT THE WAY AN ELECTRONIC FORUM WORKS

In early 1994, as part of my research at Stanford University, I undertook to investigate the way electronic forums worked by interviewing users and managers and by observing some of the ways forums were manipulated. The descriptions that follow are based on these interviews and observations, which I sought to compare insofar as possible with my own manipulations and analyses of the messages exchanged.[1] A glance at the results of more recent studies of electronic communication (and these have proliferated since the late 1990s)[2] suggests that the observations reported here remain entirely pertinent today.

---

[1] For complementary analyses, see Claude Rosental, "Les travailleurs de la preuve sur Internet: Transformations et permanences du fonctionnement de la recherche," *Actes de la recherche en sciences sociales* 134 (September 2000): 37–44.

[2] On this subject, see Claude Rosental, "Le rôle d'Internet dans l'évolution des pratiques, des formes d'organisation, et des réseaux de la recherche," *Annales des mines* (February

It quickly became obvious that newsgroups could be compared to collective mailboxes in which participants could post messages and read those sent by others. Some texts were archived, including those addressed to comp.ai.fuzzy. According to the accounts I gathered, many university faculty members and researchers in computer science, especially in the United States, were already connected to the Internet through their university or their company, and thus had access to such forums, which generally focused on specific themes. My analysis of the comp.ai.fuzzy archives confirmed this representation. More precisely, it showed that the participants in this forum were by and large members of departments of computer science in North America or researchers in computer science working for private enterprises in the United States.[3] Many of them were working on fuzzy logic.

The economy of access to electronic forums was a major factor in their use by the university faculty members and other computer scientists I encountered. Daily access to a computer connected to the Internet made it possible to interact with any electronic forum at any time in just a few seconds. Thus, at an appropriate moment, typically during a work "break" or after taking care of personal electronic mail, a researcher could easily read messages sent to a forum or post a response via this medium.

The modalities of reading and writing messages were essentially the same as those encountered in the ordinary use of a computer: the texts were typed through the use of a keyboard and read silently on screen by an individual, who could choose to save the content electronically and to print a hard copy. The reading and writing thus took place at a *workstation*—this point, although at first glance a trivial one, contributes importantly to the basic description of the practices of the theoreticians we shall meet. What some of them would describe spontaneously as proceeding from a production of abstractions was not obtained as the output of a process that was purely abstract in itself, but stemmed rather from the localized and concrete task of typing and retrieving data, manually and visually, from a computer screen.

1998): 103–8. This article opens up access in French to the large body of English-language studies of electronic communication undertaken in a first instance by specialists in communication and information science. The literature includes at least two major components: a large set of localized studies on the use of communication networks, and the thematization of the virtual in the dissertation mode. In the latter mode, a number of authors have announced the advent of an era of communications authorizing the construction of new autonomous universes that are immaterial in nature or that connect pure minds.

[3] As a general rule, the authors supplied information in their messages regarding their own identity and the institution to which they were connected. They presented themselves and their research in a variety of ways.

Moreover, the actors' electronic access to these collective mailboxes facilitated the retrieval of texts that had been saved in memory on various computers linked to the electronic networks: actors could consult not only comp.ai.fuzzy archives, for example, but also articles, theses, computer programs, and so on. Certain messages sent to comp.ai.fuzzy thus referred to texts accessible on the networks without paper support (by direct transfer or through a process known as File Transfer Protocol, or FTP); the original message included the electronic address from which the additional document could be downloaded.

For the actors, downloading texts on electronic networks was clearly preferable to other forms of access for several reasons. First of all, in just a few seconds they could capture a text and save a paperless copy on a computer. In addition, the use of electronic means of communication cost faculty members nothing, while there were costs associated with photocopying an article, sending a fax, or mailing a letter.[4]

The existence of comp.ai.fuzzy requires some clarification at this point. At the time of my study, Usenet, the network that supported comp.ai.fuzzy, classified its forums according to a tree diagram. At the base, the tree had seven branches, each encompassing a multitude of forums. For example, the branch called "comp" (an abbreviation for "computer") housed forums related to the field of computer science, such as a newsgroup on artificial intelligence called "comp.ai." But the subbranch comp.ai in turn included several more specialized forums, such as the newsgroup on fuzzy logic called "comp.ai.fuzzy."

Usenet newsgroups were devoted to a wide variety of topics, ranging from science fiction literature to "themes" associated with philosophy or artificial intelligence (for example, fuzzy logic). The number of these forums was increasing rapidly in the mid-1990s. Subbranches of the tree were created after an electronic "vote" in a preexisting forum—the branch to which the new forum might be attached. It was up to the voters to decide whether there was sufficient interest to warrant creating another newsgroup. A review of the earliest messages in the comp.ai.fuzzy archives revealed that this forum was created in January 1993 after comp.ai users had exchanged messages and then voted on the question.

---

[4] I do not propose to offer an in-depth study of the economic issue here, but it is clear that the topic would warrant much more detailed analyses, comparable to those that have been undertaken for more traditional media. See especially the studies highlighting the interest of a material history of reading presented in Roger Chartier, ed., *Les usages de l'imprimé: XVe–XIXe siècle* (Paris: Fayard, 1987). On the role of the emergence of printed material in the development of rational practices, see Elizabeth Eisenstein, *The Printing Press as an Agent of Change: Communication and Cultural Transformations in Early Modern Europe* (Cambridge: Cambridge University Press, 1972). For a history of the scriptural forms pecu-

According to the accounts I gathered, messages addressed to a given newsgroup were either placed on the forum without any form of screening or else they were censored to a greater or lesser degree by the intervention of one or more forum managers called "moderators." In the latter case, the choice of a moderator and the designation of his or her responsibilities resulted from complex processes. The decision to resort to one or more moderators might be made by vote at the time a forum was created, or it might be made later, for example (and the example is typical) in the aftermath of a scandal caused by messages containing racist remarks.

The text presenting comp.ai.fuzzy that was posted on the forum site and regularly updated thus stipulated that the discussions were "lightly moderated." There was no evidence of any real censorship, however, with one possible exception. Following a series of messages devoted to computer-based musical composition, some readers had posted messages requesting that such exchanges, deemed off the subject, no longer be pursued on the site—and discussions of this theme came to an abrupt halt.

It became clear in course of my study, furthermore, that certain forums archived all or some of the exchanges they hosted. One or more "volunteers" took responsibility for maintaining and updating the archives. This was the case for comp.ai.fuzzy. Messages addressed to this newsgroup remained in the "mailbox" for only one week before they were archived at Cornell University in an electronic database accessible "to the public," that is, to any individual connected to the electronic network.

Moreover, in the newsgroup context, "volunteers" often took charge of updating the text presenting the theme of the forum, including the "Frequently Asked Questions" (FAQ) to which users seeking to know more about the group could refer. Comp.ai.fuzzy had a regularly updated FAQ section that focused particularly on the "theory" and "applications" of fuzzy logic.

These structural descriptions should suffice to set the stage for the analyses that follow. However, before we look into the controversy that unfolded on comp.ai.fuzzy, it will be useful to evoke some of the results of my fieldwork on the teaching of logic in a major American university, in order to bring to light the vocabulary and skills shared by many of the actors in the debates studied in this book.

A look at some of the serious difficulties encountered by academics in the teaching of logic provides a good vantage point for analyzing practices that quickly become opaque when they become matters of routine, for introducing elements of a descriptive language adapted to the phenomena at issue, and for showing how the production of assertions in logic can

---

liar to mathematics (from 1300 to 1600), see Cynthia Hay, ed., *Mathematics from Manuscript to Print: 1300–1600* (Oxford: Clarendon Press, 1988).

give rise to debate. The analysis should thus make the subsequent developments easier to follow for readers who may be completely unfamiliar with
logic and also for scholars who may not deal with logic in the same way
as those observed.

SHARED SKILLS IN LOGIC

In 1992, seeking a grasp of the way logic was taught in the United States,[5]
I "followed" with particular attention a course in logic offered by the
philosophy department in a major American university. I observed lectures, discussion groups, paper grading, and interactions that took place
during office hours between students, the professor in charge of the
course, and the teaching assistant responsible for the discussion groups.
A series of interviews with the various protagonists completed the study.

Within the philosophy department, this particular course was conceived as part of the university's "general education" program. It was
taken by roughly one hundred students of all levels—primarily students
who had chosen to major in philosophy, mathematics, or computer science. According to philosophy department faculty members, the course
was designed to give the students who took it some "basic notions" in
formal logic. It was different from other, optional courses in logic that
were viewed as offering much more specialized knowledge in the field.
This basic course was moreover a follow-up to a course devoted to so-
called "baby logic," that is, to the teaching of syllogisms in prose.

An examination of interviews I conducted with faculty members in this
and other North American universities, archival data about the evolution
of university curricula,[6] and several editions of logic textbooks made it
clear that the content of basic logic courses (which are very widespread
in the United States) has been remarkably stable since the end of World
War II.[7] The goal of such courses is to give the students both a language
and a fairly large array of skills to use in symbolic manipulation.

[5] For a more detailed presentation, see Rosental, "De la logique au pathologique." See
also Claude Rosental, "An Ethnography of the Teaching of Logic," paper delivered at a
colloquium titled "Historical Epistemology," Institute for the History and Philosophy of
Science and Technology, University of Toronto, November 1993.

[6] The annual course catalogs in which universities publish their courses of study were
valuable sources in this regard.

[7] Examination of course catalogues in the Harvard University archives indicated that
courses in logic had been introduced early in the twentieth century and had gradually been
substituted for courses in rhetoric, courses that had themselves been destined initially for a
very broad student public.

The instructors aimed first of all at constituting "logic" as a separate and autonomous *language.* To this end, certain words attributed to "ordinary" mathematics were retained and consecrated as *logical* terms par excellence. These words were characterized as logical *symbols,* and distinguished from other words that were said to belong to "ordinary language."

Visual supports were brought in to shore up this distinction. Thus the words "and," "or," and "not," called connectors because they make it possible to link statements together, were *symbolized* by "∧", "∨", and "⌐" respectively. The terms "implies" and "equivalent" were represented respectively by the signs "⇒" and "⇔". The expressions[8] "for all" and "there exists," called universal and existential quantifiers, were similarly *symbolized* by specific graphic signs "∀" and "∃". Capital letters were generally used to *symbolize* any sort of statement. The letters "P" and "Q" were often used to represent such *predicates* or *properties.* "P", for example, could be used as an *abbreviation* of the statement or else of the property "to be a man." The statement "*x* is a man" could thus be represented by "*x* has the property P", or, more rapidly, by adding the small letter *x* immediately after the capital letter P: "P*x*".

Starting from certain dedicated words and letters that could be used to represent "everything else," the instructors announced to the students that they now had in hand a "new" language, called a formal language, and that the language in which they had been expressing themselves up to that point was "ordinary" language. One of the advantages of the formal language would be its "precision," or its "rigor," whereas everyday language was presented as a source of "confusion." To reinforce the opposition between these two languages, the instructors would ask the students to *practice,* that is, to carry out a certain number of manipulations—in particular, to *translate* propositions from "ordinary language" into this "new language" and in that sense to express themselves in it.

The students had to translate, for example, the proposition "John passed the test or John failed it." To *symbolize* this statement, the students had to try to bring logical symbols such as those we have just seen into play. Little by little, they had to acquire the skills corresponding to the following series of tasks: identify the connector "or" that divides the sentence in two, and represent it by "∨"; identify the predicate "John passed the test" and represent it by a letter, for instance "P"; then identify the negation of this statement in the second part of the sentence, "John failed it," and represent it by "⌐ P"; finally, propose the symbolization "P ∨ ⌐ P".

---

[8] The words "term" and "expression" are themselves dedicated terms in logic, but I am using them here in a nontechnical way.

The students were given to understand that by carrying out such a task they were revealing the "formal structure" of the sentence. The instructors were thus affirming that a "hidden structure," the "form" of the statement, was to be found behind the sentence expressed in prose, and that "logic" constituted precisely the language by means of which that structure could be revealed.

As the students progressed, the work of identification and translation increasingly entailed a search for *paraphrases*. In order to unveil the formal structure—that is, to bring certain "underlying" logical terms to the "surface"—of the sentence "Every professor who is an administrator earns more than certain professors who are not administrators," the exercise of formalization required that the students be able to construct the following intermediary paraphrase: "For every individual x, if x is a professor and x is an administrator, then there exists an individual y such that y is a professor and y is not an administrator, and y earns less than x."

This specific work of translation, centered on the search for a display of certain words assigned to the vocabulary of logic, was also constitutive of a whole series of dichotomies. The translation process consecrated the oppositions between formal language and ordinary language, between precision and vagueness, between what would be "formal" and what would be "intuitive" or "contextual."

In the end (and this may strike some readers as "counterintuitive"), the formalization of statements stemmed, as such, from a *material* practice of writing. In this regard, it should be noted that students usually found it quite difficult to associate the notion of formalism with the operation that consisted in *bringing into play* the small set of words that they often viewed as arbitrarily chosen, despite their instructors' argument that the terms were in common usage in mathematics. To develop these points, an illustration based on student-teacher exchanges may be useful.

The first exchange presented below took place during a discussion group session between two students and a teaching assistant (represented here as Student A, Student B, and TA). The conversation followed the resolution of a problem-solving exercise put on the blackboard by the TA; the task was to formalize the expression "the current president." The solution proposed by the TA consisted in translating the expression by the following series of paraphrases, intended to bring "logical" terms into play in stages (to make the modalities of this process more apparent, certain words appear in bold type):

1. **There is** a current president and **there is not** more than one president.
2. **There is** a president, and **if there is** another president, **then** it is the same one.

3. **There is** an x, such that x is president, and **if there is** a y, such that y is president, then x **equals** y.
4. **There is** an x, **Px, and if there is** a y, **Py, then** $x = y$ [with "Px" being an abbreviation of "x is a president"].
5. $(\exists x \ Px) \wedge (\exists \ y \ Py) \Rightarrow x = y$.

Once this written work had been completed at the blackboard, a student asked the TA why it was necessary to specify that there was not more than one president:

STUDENT A: Why do you need that? There can't be more than one president of the United States.

TA: Well, because it is not really clear from . . . this is something that you assume, but it is not perfectly determined by this model, uh, that there is only one.

STUDENT A: OK, OK. Outside knowledge does not apply?

TA: Right. [Silence.] OK, uh . . .

STUDENT B: So in that case, you don't have to define what a president is, do you?

TA: Pardon?

STUDENT B: If you don't apply external knowledge, don't you have to define, then, what a president is?

TA: OK, I mean, of course you have to know what you mean by the words. But you don't have to know that there is only one president, or something like that.

STUDENT A: But I mean, when you are getting an arbitrary model, you know, if you use as universe of discourse "People of the United States of America," there can only be one president, and that is not something that is external to the concept as . . . If you don't accept that as good enough, then you can't accept the knowledge of what a president is, because it is going to be ambiguous as, like the president being unique in the United States.

TA: Well, let's see . . . Actually, yeah, I mean, you have to construct symbolization keys so that you don't even have to use any knowledge of . . . You don't have to know . . . Of course you have to know something, you have to know what is "a," you have to know what this . . . But this is a distinction between, uh, the language of logic and the language of metalogic. We use, to explain something, we use, uh, I don't use the language of logic. Because there is nothing like . . . So I use the "is a," right, that means that you understand. But for saying "current President," I don't have to know what is the current president. [Silence.]

Reading this exchange allows us to grasp the nature of a persistent misunderstanding between students and instructors concerning the notion of formalism, a misunderstanding that the TA's improvised response (which would normally have been more "efficient") did not succeed in overcoming here; instead, it left some of the students (and perhaps also the "novice" reader, that is, the reader unfamiliar with formal logic) even more confused. At the same time, reading the exchange allows us to see clearly what the academics meant by formal knowledge. The students were in fact learning only gradually, at best, through exercises like this one, to attribute the same meaning as their professors to the opposition between formal and informal knowledge.

For the TA, the series of sentences on the blackboard constituted a "logical analysis" or a "formalization" of the expression "the current president" inasmuch as dedicated logical symbols were *introduced*, in a way corresponding to a certain practical skill. In particular, for the TA, the expression of a unicity clause ("there is not more than one") in a paraphrase presented as equivalent to the first expression was a classic way of revealing the formal structure of the expression, for it offered the opportunity—one with which he was very familiar—to *bring* standard logical symbols *to light* at a later stage.

Some of the students did not (yet) realize that the term "formal" was synonymous, *for the instructors*, with the deployment of this specific know-how. For the students, "formal" seemed to mean something like "intrinsic knowledge," as opposed to "external" or "contextual" or even "ambiguous knowledge." To the extent that they did not share the instructors' understanding of the dichotomy between formal and informal knowledge, the students found the expression of a unicity clause superfluous. For them, the fact that there is only one president of the United States constituted an *already* formal bit of data. Under these conditions, what led the instructors to qualify certain notions as formal or contextual, and to emphasize the need to formalize certain statements, struck many students as totally *arbitrary*.

More broadly speaking, the instructors sometimes had great difficulty convincing the students that this sort of work was worthwhile. Some students were openly skeptical about the usefulness of logic in general and of formalization in particular. Their teachers often used student comments or questions along these lines to stress certain dichotomies, and in particular to insist on the thesis according to which ordinary language, unlike formal language, is ambiguous. This was precisely the reason, they sometimes asserted, why logic was actually of limited value for the analysis of ordinary language. When they beat a retreat in this way, the instructors stressed the fact that, in contrast, formal logic was of the greatest value for the expression of mathematical assertions, and these, the instructors declared, were exempt from any ambiguity. Thus during one lecture, after the professor had had a great deal of trouble convincing the students of the value of a specific formalization, he declared:

Natural language is not formal. The interpretation of "if" depends on the context. If I say "John," many people will stand up. Sometimes, "if" means "if and only if"; it depends on the context. Modern logic was founded for mathematics, and in mathematics, you don't have any ambiguity. . . . We don't really have a perfect symbolization of all English sentences in this language. What we do have is—we have a perfect symbolization of all mathematical statements in this language. And that's of great importance. So while logic, formal logic, is a perfect or a very good language for mathematics, it is also a good language for general thoughts, general inference, but it is not a perfect language. OK? Still a lot is left out. OK? [Silence in the classroom. The professor then goes on to another topic.]

To support declarations like these, the instructors showed the students how they could formalize certain mathematical assertions. But even on these occasions they sometimes came up against overt skepticism, or, more often, polite or mocking irony on the part of the students concerning the value of such an exercise. Let us examine one of these attempts and the reactions it produced in order to get a better sense of the nature of the approach in question.

During a lecture, a professor was explaining how formal logic made it possible to express the very fact of counting, without using numbers. As an "example," he proposed to formalize the phenomenon of counting to three. "To that end," he showed the students how to formalize the assertion "There are at least three students in the class." He began by formulating a *paraphrase* presented as equivalent to that assertion:

> There is an x, there is a y, there is a z, such that x is in the class and y is in the class and z is in the class, and x is not equal to y, and x is not equal to z, and y is not equal to z.

This paraphrase allowed him to go on and *introduce* logical symbols and to propose the following symbolization (using "Px" as an abbreviation of "x is in the class"):

$$\exists x\, \exists y\, \exists z\, (Px \wedge Py \wedge Pz) \wedge (\neg(x = y) \wedge \neg(x = z) \wedge \neg(y = z)).$$

Not without irony, a student then asked the following question:

STUDENT: How would you say there are thirty students?
PROFESSOR: Oh! [Students laugh for a few seconds.] I could say there are thirty students, but it would probably take the whole hour. OK? [Students laugh again for a few seconds.] I would have to say: "There is an $x1$, there is an $x2$, an $x3$, until $x30$, and the list of inequalities that are all implied."

In raising his question, the mischievous student was obviously seeking to make fun of the cumbersome formalization proposed for such a simple statement, and perhaps to emphasize that, on the contrary, formal logic is not interesting as a way of expressing mathematical assertions.[9] Thus in front of his classmates he abruptly managed to challenge the exemplary character of the professor's patient demonstration. The question was all the more damaging, moreover, in that it led the professor to use the numbers 1 to 30 to "index" the elements introduced into his new formalization, whereas he had just declared that it was not necessary to resort to numbers in order to express the phenomenon of counting formally. In retrospect, the fact of having mobilized a lexicographic (alphabetic) order ("x, y, z") to express the existence of these three elements might well have appeared as a disguised way of counting, despite the professor's declaration to the contrary.

Thus the process of establishing the usefulness of formal logic for the analysis of ordinary language and for the expression of mathematical assertions posed numerous difficulties for the instructors. For them, this process required major efforts (while for us it represents a tool that can give us a better grasp of the logical practices in play). The same thing can be said, as we have seen, about the process of conveying a specific meaning to major dichotomies such as the one between formal and informal knowledge, a meaning that was best inculcated through practical exercises.

These specific difficulties were compounded by the fact that controversies kept breaking out between students and faculty about the correctness of the various formal translations proposed during the exercises. These exchanges, sometimes worthy of Ionesco,[10] occasionally led to open conflict, especially when student papers were handed back and corrected at the blackboard. For a given exercise, many competing translations were often produced, and the legitimacy of one proposed translation was often contested by the authors of others. Analyzing this phenomenon helped bring to light the considerable number of tacit practices that had to be acquired by students so they could produce the "equivalent" paraphrases that the instructors were looking for.

[9] On this topic, see the criticisms leveled by Philip J. Davis and Reuben Hersh mentioned above (chapter 1, pp. 40–41). Let us recall that, according to these two authors, formal logic is a poor language for expressing mathematics: couched in such "cumbersome" terms, mathematical demonstrations would generally be obscure and unverifiable. This handicap is compounded by uncertainties as to the correctness of the translations carried out and by the sometimes insurmountable difficulties that stand in the way of achieving such translations either "in principle" or "in practice." See Davis and Hersh, "Rhetoric and Mathematics."

[10] See Eugene Ionesco, "The Lesson," in *Four Plays*, trans. Donald M. Allen (New York: Grove Press, 1958), pp. 43–78.

This chaotic situation emerged in spite of the instructors' use of *step-by-step questioning* and sophisticated strategies of visual *identification* to guide the students through the work of translation when they were solving problems at the blackboard. The instructors showed the students how to carry out canonical translations by asking, for example, where connectors, quantifiers, or predicates "were located." At the board the instructors sometimes underlined connectors, boxed in predicates, and added subtitles to indicate the "presence" of a universal quantifier, for example by writing "for all" under the term "each."

During a lecture, for example, in order to "extract the formal structure" of the two sentences "If you love me, I will marry you" and "I will marry you only if you love me," and in order to demonstrate that the two possess the same form, the professor put the following inscriptions on the board:

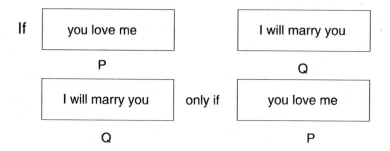

By framing certain word groups constituted as predicates, then labeling them with the letters P and Q and thereby isolating the conditional logical symbol, the professor *showed* that the two sentences had "in fact" the same "form": "$P \Rightarrow Q$". Similarly, during the discussion groups, by teaching them "tricks," the TA *trained* the students to *see how to bring* the formal structure of statements *to light*. He tried especially to inculcate questions that would become automatic, and he used various markers to underline the words that made it possible to introduce logical symbols, as in the following exchanges:

TA: Let's start with "Charles is rich and beautiful, but not intelligent." How do you symbolize it? What is the main operator?

A STUDENT: It's the "but."

TA: Right, very often it comes after the comma. You have "blah-blah-blah" and "blah-blah-blah." [TA writes on the blackboard: $(\ldots) \wedge (\ldots)$.]

TA: Let's do exercise J. Again, think first what is the main operator. You have "but." "But" is always translated by "and". You have something like this:

"blah-blah-blah" and "Charles is" [TA writes on the blackboard: (. . .) ∧ (Charles is)].

TA: How do you symbolize "Either all are black, or all are red"? The things that are important are "either," "all," "or," "all." [The TA underlines the words at the blackboard: either all are black, or all are red.]

Teaching the "logical analysis" of language thus consisted in teaching students to recognize what could be viewed as constituting the "formal structure" of statements. It relied on the use of markers in the context of written exercises. In other words, teaching students how to perform the cognitive act that consisted in *seeing how to bring* the formal structure of a statement *to light* was inseparably linked to the use of elaborate graphic and visual technologies.[11] The use of these technologies corresponded to authentic pedagogical *skill*. To this end, the instructors had identified sources of standard "good" examples and "good" exercises.[12] The teaching of logic was thus not carried out along the lines of a philosophical dissertation, a conceptual history, or an oral exchange comparable to those in Plato's dialogues. It consisted above all in an attempt to transmit to students graphic and visual *skills* and *automatic reflexes* that were coming to characterize the emerging exercise of logic.

However, the students also had to practice manipulations other than those implied by the exercise of formalization (these others, too, gave rise to many controversies). In particular, the process of learning "propositional calculus" involved new practices of reading and writing. This "calculus" consisted in attributing a "truth value" (1 for true, 0 for false) to a complex proposition obtained by using logical terms to combine simple propositions whose truth values were known or fixed.

One task, for instance, was to define the truth value of "P ∧ (Q ∨ R)" in the case in which P is true (i.e., has a truth value of 1), Q is false, and R is true. To do this, it was necessary to begin by defining the "action" of the connectors "and" and "or," that is, to establish the truth value of the propositions "A ∧ B" and "A ∨ B" as a *function* of the truth values of A and B (for example, if A has the truth value 1 and B has the truth value 1, then A ∧ B has the truth value 1). Since each proposition could have only two truth values, 1 or 0, there were only four possibilities for each connector, and this function was represented in a chart known as a "truth table."

---

[11] See also Hutchins's "technological" analyses of the processes of learning and accomplishing tasks that put formalisms into play: Hutchins, *Cognition in the Wild*.

[12] The nature of the pedagogical skill in question was suggested in remarks like the following, formulated by the TA during one of the discussion groups: "By the way, families and numbers are always good examples."

A	B	A ∧ B
1	0	0
1	1	1
0	1	0
0	0	0

Let us note here that, to the extent that propositions could have only two truth values, the logic involved was called binary, classical, or Boolean,[13] as opposed to a "multivalent" and "nonclassical" logic that authorizes calculations on the basis of a larger number of truth values. From charts like the one above, the "calculation" of the truth value of more complex propositions could then be obtained by constructing larger charts through the use of skills best inculcated through *repetition*.

The procedures for constructing this propositional calculus served in particular to "demonstrate" "classical logical principles." The principle of the excluded middle (according to which "A proposition is either true or false") was "demonstrated" by a translation: the truth value of "P ∨ ¬ P" is 1 whatever the truth value of P. The principle of distributivity of "and" in relation to "or" was demonstrated and defined simultaneously by the following translation: the truth value of P ∧ (Q ∨ R) is the same as that of (P ∧ Q) ∨ (P ∧ R), whatever the truth values of P, Q, and R.

As we have already seen, propositional calculus constituted only one source of the manipulatory practices introduced by the instructors. The latter also presented other ways of doing logic, especially with the help of "axiomatic" formulas. In this case, it was a matter of preparing a list of resources and a list of authorized symbolic manipulations, the mobilization of which required careful visual attention on the students' part; in turn, this kind of attention required a lot of practice.

Starting from axioms such as the principle of the excluded middle and from "deduction" procedures (that is, procedures for the development and translation of statements such as the principle of distributivity of "and" with respect to "or"), new assertions known as "theorems" were produced. Certain sets of resources, presented as corresponding to ordinary mathematical manipulations, were called "algebraic structures." For example, "Boolean algebra" constituted a particular configuration

---

[13] The mathematician George Boole was credited with establishing the calculus of logical arguments in the form of operations using the numbers 0 and 1: see George Boole, *The Laws of Thought* (New York: Dover, 1958 [1854]).

attributed to manipulations of the numbers 0 and 1 as used in computer calculations.

It is important to note that, although the instructors associated logical formalism with the notion of precision, the symbolic procedures they demonstrated were often a source of confusion for the students and led to many random results. To try to overcome these difficulties, which were compounded by the fact that the acquisition and display of skills had to take place under quite constraining temporal conditions,[14] the students had above all to make the *effort* to practice solving typical problems like the ones illustrated above, and to do so *repeatedly*. This discipline often allowed them to progress relatively quickly in acquiring the *tacit practices* they needed so they could deal with the random elements they encountered during the "application" of formal procedures.

This dimension must be taken into account if we are to understand what logical work properly speaking might entail. The repeated practice of canonical exercises corresponded quite closely, in the end, to certain methods for learning to play a musical instrument such as the violin or the piano, entailing the frequent practice of scales.[15] Moreover, in both cases *manual* work is involved (moving one's fingers "methodically" in practicing an instrument, carrying out writing tasks with the help of a pen in practicing logic), along with an effort to *visualize* (reading scores or watching one's fingers in order to play an instrument, *seeing* how to *bring to light* logical symbols or a series of equivalent statements in order to complete a logic exercise). In each case, the skill can be acquired through persistence,[16] but the same skill (and especially a certain "agility") can also be partly lost over time owing to lack of exercise. By the same token, moreover, while it is possible to describe how logic is taught and practiced in certain places, it is much harder to transmit "skill" in the area of symbolic manipulations to "novice" readers (to that end, readers would not only have to *read* this text but they would also have to *practice*).

The instructors presented this mode of learning logic to the students as the only way, or at least as the most efficient way, to overcome the significant difficulties the latter often faced. It must be noted in this connection that, despite the considerable ingenuity manifested by the instructors in developing strategies for "inflating" grade averages and for "rescuing"

[14] Learning had to take place within a limited time frame (a few months) and the students had to be able to show their competence on examinations for which specific time limits were set (for example, several exercises had to be completed in the space of an hour).

[15] For a sociological study dealing with modes of learning music, see Antoine Hennion, *Comment la musique vient aux enfants: Une anthropologie de l'enseignement musical* (Paris: Anthropos, 1988).

[16] In the case studied by Hutchins (*Cognition in the Wild*), too, the actors required much less time to carry out the tasks themselves than to learn to do them correctly.

the weaker students, the failure rate[17] in this course reached a record level in comparison to other courses offered at the university in question, matching the level traditionally attained in the introductory course in organic chemistry. Here is an indicator that highlights once again—indirectly and in an aggregate fashion, to be sure—the considerable number of tacit practices that the students had to acquire in order to succeed in reproducing the manipulations advocated by the instructors. The failure rate was also related to the multiple sources of dissension that emerged in the exercise of logic.

In the face of the students' poor results on the tests or midterms that punctuated the course, the instructors constantly urged the apprentice logicians to attend all the discussion groups and to go over the exercises that had been presented in that context. They sometimes even went so far as to prepare and transmit exhaustive lists of definitions and exercises on which the students would be questioned during tests and exams; their intent, they said, was to "prepare" the candidates better. Thus the professor stated during one lecture:

> You will actually go over all types of exercises that I will give in the midterm in the discussion groups, and we encourage you very much to go to all the discussion groups, pay attention to what is going on in class, definite descriptions we did in class, the definition of being a cousin we did in class. So all these things we did in class, the questions about models, the two sentences, whether they are true or false in a given model, exactly the same questions the TA did in the discussion groups. So that practically, all the questions that will be in the exam will either be done in class at the blackboard or in the discussion groups. So I urge you very much, first to pay attention to the examples that I do on the blackboard—for sure, they should reappear in the exam. Second, to go to the discussion groups, and pay attention to the exercises the TA does over there. Again, they are very likely to be in the exam. As for the definitions I gave you, I will give you a handout with all the definitions that I expect you to know. OK?

In addition to preparing students for tests, another crucial dimension of the instructors' efforts to train students to manipulate logical symbols needs mentioning in order to complete this analysis: the *accompaniment* of formal statements. My use of the term "accompaniment" refers to a set of mediations mobilized by the instructors in an effort to legitimize, in the students' eyes, specific modes for grasping and producing statements, in connection with establishing equivalences or using deduction

---

[17] Here I am referring to the relation between the number of students who failed to pass this course and the number of students who were registered in the course on the day of the final examination (at the beginning of the course, the number was significantly higher: in other words, the discouragement rate was also quite high). For precise statistics, see Rosental, "De la logique au pathologique."

procedures. These mediations sometimes stemmed from appeals to *intuition* or to *self-evidence*, with the help of images, discourses, and all sorts of references to forms of life, to borrow Wittgenstein's vocabulary. In particular, brief scenarios might be used to dramatize behaviors from "everyday life," scenarios intended to justify the choice of certain formal manipulations.[18] The following examples will serve to illustrate this point.

During a lecture, in order to justify taking as true an assertion of the form "P⇒Q" in which the premise "P" is false, the professor called on the students' "intuition" in the following terms:

> Let me give you an example that will bring you at least intuition of it: "If you love me, I will marry you." Now, if you don't love me, but you marry me, it is clear I did not lie. What I say is consistent with what I said, so we take it to be true. It is our convention and we will take our intuition to take it as a convention.

We see from this passage how the professor was led to define a realm of intuition in order to constitute that of formalism. The opposition between intuition and formalism played an essential role in the construction of each of these two fields. Intuition appeared as an initial moment that did not possess the same degree of solidity as the moment proper to symbolic manipulations. Its role was essentially to provide access to the meaning of the manipulations and to their correct execution. Although things were generally not this simple in the exercise of logic, this declaration suggested that while the "good reasons" of intuition were useful in a first stage, they could nevertheless be left behind later on when "purely" symbolic manipulations were at stake.

Sometimes, if only to give meaning to certain statements, the instructors resorted to constructing diagrams, presenting them as "supports" for intuition. When a proposition expressed through logical symbols was juxtaposed with the same proposition expressed through a figure, a new meaning for the statement was supposed to emerge, the meaning that the instructors themselves attributed to it in their commentaries. For example, to give meaning to the correctness of the proposition "$\forall x (Px \lor Qx)$," during a lecture, the professor drew the following figure:

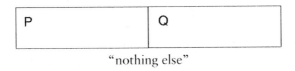

"nothing else"

---

[18] Of course the role of such scenarios in modeling the social world can be observed in many other areas. See especially Sabine Chalvon-Demersay, "Une société élective: Scénarios pour un mode de relations choisies," *Terrain* 27 (Sept. 1996): 81–100.

In constructing such a diagram, the students were supposed to have the "intuition," as the instructor spelled it out in words, that the universe in which this proposition was true was limited to the combination of these two sets (the one for which the property "P" was verified and the one for which the property "Q" was verified).

Sometimes, too, the instructors legitimized a formal manipulation not by appealing to the students' "intuition," but by emphasizing the "self-evidence" of the manipulation. For example, the equivalence between the statements "$\neg \ (\forall x Px)$" and "$\exists x \ \neg \ Px$", written on the blackboard during a discussion group, was presented by the TA in the following terms:

> Of course, you just have to take a look at these forms for two or three minutes, and you'll see why they are equivalent. The only thing that you have to do here is simply to change the quantifier and to put negation. It's no problem.

For the TA, the equivalence between these two propositions was thus "self-evident," or "no problem." According to him, just a glance (however prolonged) sufficed to *bring into view* the link (of equivalence) that connected these two statements. The "glance" in question consisted in assuring oneself that the relevant skill had been properly put to work, by verifying that, from one proposition to the other, the displacement of the symbol of negation toward the right was accompanied by a change in the quantifier. Beyond the role played by acts of visual verification in the perception of the correctness of a logical statement and in particular of such an equivalence, the double performative[19] function of this discourse on self-evidence needs to be noted here: the process in the course of which that equivalence was *said* to be self-evident helped make the manipulation in question acceptable, on the one hand, and helped constitute a realm of self-evidence in the students' eyes, on the other hand. The equivalence was thus *made self-evident*.

In the final analysis, the constitution of formalism passed in a wholly fundamental way through the constitution of the realms of self-evidence and of intuition. This latter was constituted, mobilized, and sometimes even adjusted to give a meaning to so-called formal manipulations and to justify them.

However, we must note that, in practice, demonstrations were virtually never limited to a single register. They entailed constant comings and goings between discourses on self-evidence, intuitive justifications expressed

---

[19] Performative or constitutive, in the very act of enunciation. On the question of the primacy of public expression for grasping speech acts, see Louis Quéré's analyses in "Agir dans l'espace public: L'intentionnalité des actions comme phénomène social," in *Les formes de l'action*, ed. Patrick Pharo and Louis Quéré (Paris: Éditions de l'École des Hautes Études en Sciences Sociales, 1990), pp. 85–112.

orally or with the help of diagrams, and symbolic manipulations. In other words, even when, as if in response to critics such as Lakatos,[20] Davis, or Hersh,[21] the instructors attributed the necessity of the demonstrations to formalism alone, the formal "autonomy" of the demonstrations was in fact a form of freedom under surveillance.

We need not pursue this description any further: the practices and vocabulary we have been examining should suffice to allow us to navigate relatively easily from here on in the debates we are about to study.

At this point, it will be helpful to examine the messages that were proffered on the electronic forum comp.ai.fuzzy in immediate response to the publication of Elkan's text. Beginning with an overview of the protagonists' interventions will make it possible to develop more fully the descriptive tools that have already been introduced. After presenting the data in this way, I propose to analyze the specific *dynamics* characteristic of the debates, with a special emphasis on the way a certain number of irreversibilities, particularly positions that are *stronger* than others (in a sense that remains to be specified) manage to emerge in the course of the exchanges.

In this connection, it may be worth emphasizing that, in the next chapter but also more generally throughout the entire book, the actors' work will be depicted with increasing precision. I shall bring to light more and more distinct mediations of logical activity: indeed, it is because they are so numerous and because they cannot be reduced to one another that I can deploy them only gradually. Thus, even if the reader has the impression of having "covered the territory" at the end of a given section, he or she must expect that, until the very end, we shall move, so to speak, from one surprise to another. In other words, and to conclude on a cautionary note, let me stress that the actors' work I am about to describe cannot be fully grasped on the basis of a partial reading of this book; in no case can it be reduced to the emerging representation of just one facet of the analyses presented here.

[20] Lakatos, *Proofs and Refutations.*
[21] Davis and Hersh, "Rhetoric and Mathematics."

PART TWO

PRACTICES OF DE-MONSTRATION: DEBATING A THEOREM IN AN ELECTRONIC FORUM

"Were you at Eliane de Montmorency's yesterday, cousin?" ... "Good gracious, no; I like Eliane, but I never can understand her invitations. I must be very stupid, I'm afraid. ... Above the questionably authentic name of 'Montmorency' was the following kind invitation: 'My dear cousin, will you please remember me next Friday at half-past nine.' ... Although renowned for my obedient, punctual and meek nature, ... I was a few minutes late. ... at the stroke of ten in a comfortable dressing-gown, with warm slippers on my feet, I sat down in my chimney corner to remember Eliane as she had asked me and with a concentration which began to relax only at half-past ten. Tell her please that I complied strictly with her audacious request. I am sure she will be gratified."

 —Marcel Proust, *The Captive*, trans. C. K. Scott Moncrieff, in
 Remembrance of Things Past, vol. 2 (New York: Random House,
 1927), p. 565.

Chapter 3

BRINGING TO LIGHT: DEMONSTRATION PUT TO THE TEST OF ANTAGONISTIC LOGICAL PRACTICES

THE FORMAL PRESENTATION OF A DEMONSTRATION DOES
NOT LEAD AUTOMATICALLY TO CONSENSUS

*The Absence of Universally Recognized Central
Logical Principles*

ELKAN'S PROOF was published in July 1993 in the proceedings of the AAAI'93 conference, and strong reactions began to appear very quickly on the electronic forum comp.ai.fuzzy. For a year, starting in August 1993, and especially during the first three months, participants in the forum sent a great many messages on the subject. These messages were accessible on the forum for a week before they were archived. While they varied in length from a few lines to a few pages, on average they ran to several paragraphs.

Given the nature of comp.ai.fuzzy, which amounted to a sort of collective mailbox, as we have seen, the messages were primarily addressed to a group of anonymous readers who were likely to connect to the forum. But they were sometimes addressed at the same time to authors of earlier messages, in which case they were presented as responses or reactions. Inasmuch as many writers set forth their own points of view while citing excerpts from earlier contributions, the texts generally proceeded from a complex interweaving of arguments, and were structured by several levels of citation. We must note that, for reasons having to do with the software used in drafting the messages, the citations were not set apart by quotation marks but by the sign "more than" (>), placed at the beginning of a line.

Systematic exploration of the comp.ai.fuzzy archives over a two-year period, from its creation in January 1993 to the end of 1994, allowed me to identify the messages that followed the publication of Elkan's article. I was able to access these archives by transferring them electronically from the database maintained at Cornell University in Ithaca, New York, to a computer at the École des Mines in Paris, and then to diskettes in order to analyze their contents on a personal computer.

The messages sent to the comp.ai.fuzzy newsgroup were automatically arranged in chronological order. Sequential numbers were attributed to them as they were received on the forum. The absence of certain numbers thus attested to the incomplete character of the archives;[1] however, this was a relatively limited phenomenon. In fact, the frequent citation of messages in other texts compensated for the absence of certain contributions in their original version. Moreover, every previous submission referenced but not cited in a given message could be found, at least in part, in a third archived intervention. In other words, although certain messages were completely absent from the archives, these were neither quoted nor referred to in any of the messages present in the corpus, so that my data formed a connected set.

The volume of textual production was quite high, owing primarily to the practice of citing earlier messages, and also to the numerous and very lengthy announcements of conferences that were addressed, sometimes repeatedly, to the newsgroup. From this data set, all the messages that followed upon the publication of Elkan's article could be excerpted for detailed analysis. In order to follow the debates, it was indispensable, in addition, to consult texts located in other electronic databases to which certain messages referred. The first qualifier that seems appropriate to describe the corpus in question is not so much "polyphonic" as "cacophonous": the term is well suited to characterize the magma of unorchestrated interventions that confronted the analyst (let us recall that this forum was "weakly moderated").

It will be useful to examine certain excerpts from these messages in order to measure at the outset the extent of the divergencies they express. While the tone generally used to refer to Elkan's formal proof[2] is one of certainty, the fact remains that the positions adopted toward it differ radically, as much on the question of its correctness—whenever this correctness is doubted—as on the nature of the errors the author is said to have made. The juxtaposition of viewpoints that follows offers a glimpse of these flagrant differences:

> From only the standard postulates of fuzzy logic . . . [Charles Elkan] proves the theorem that: For . . . two assertions A and B, either $t(B) = t(A)$ or $t(B) = 1 - t(A)$.[3]

---

[1] There is no indication that the archive's managers made any deliberate choices among the messages; the omissions probably resulted from simple negligence.

[2] The proof can be called "formal" inasmuch as it mobilizes a mathematical formalism, one that is expressed, moreover, with the help of "ordinary" symbolic inscriptions. We shall return to the details of this proof later on.

[3] From here on, the numbers attributed to the messages from which excerpts are cited will be noted systematically, using the comp.ai.fuzzy classification system. In contrast, it

Fuzzy logic is intended to be a generalization of standard binary logic. The point of my theorem is that unfortunately, this generalization fails mathematically and collapses to standard binary logic.[4]

It appears he [Charles Elkan] made exactly the mistake that I thought he had.[5]

Elkan needs the law of excluded middle to show the logical equivalence of his implication sentences. . . . I tried his proof with (not A) ∨ B instead of B ∨ ((not A) and (not B)). I got a different result. Try it out.[6]

If there is a consistent definition of "and" and "or" in Fuzzy Logic that allows the and/or distributive law, then this supposed disproof of Fuzzy Logic fails.[7]

Zadeh's logic[8] is distributive and, therefore, the roots of Elkan's mistake do not lie on distributivity but on failure of the law of the excluded middle (as correctly pointed out by Kroger). . . . his theorem is based on incorrect assumptions, his paradox non existent.[9]

The problematic step is the first one, the use of the last axiom. I don't see the excluded middle coming in here. Whether the use of the last axiom is justified depends indeed on the interpretation of logically equivalent. If "logically equivalent" means logically equivalent in the traditional boolean sense, then Elkan is right. He correctly applies all axioms and arrives at the only possible solution.[10]

Elkan's mistake was in supposing Fuzzy Logic to be an extension of the propositional calculus in which classical stuff like LEM [law of excluded middle] would hold, instead of being a fully-fledged "deviant logic" (to borrow a phrase from Susan Haack).[11]

---

does not seem necessary to recall each time that the messages come from this specific forum; that part of the reference will remain implicit. The present excerpt is from message 788, by F. Z. Brill. It should be noted that the comp.ai.fuzzy archives could be consulted electronically by a File Transfer Protocol (FTP) command, at the following address: ftp.cs.cmu.edu/. The archives were placed in the subcategory called user/ai/pubs/news/comp.ai.fuzzy.

[4] C. Elkan, message 794.

[5] J. Wiegand, message 832.

[6] B. Dalton, intervention dated August 31, 1993, cited in message 1823. As noted above, certain original messages were missing from the archives, including the one from which this excerpt is taken. From here on, in such cases the reference number of the message in which the missing intervention was reproduced for the first time will be noted by default.

[7] L. Petrich, message 804.

[8] L. A. Zadeh was widely considered to be the "inventor" of fuzzy logic. One of his articles from the mid-1960s was usually cited in support of this affirmation: L. A. Zadeh, "Fuzzy Sets," *Information and Control* 8 (1965): 338–53.

[9] E. Ruspini, message 816.

[10] H. Lucke, message 827.

[11] P. Jackson, September 3, 1993, cited in message 1405.

Elkan's proof consists of taking a statement that is an identity for crisp logic, which satisfies the and/or distributive property just shown, turning it into a fuzzy-logic statement with definitions of "and" and "or" that do *not* satisfy this property, and finding (not surprisingly) constraints on the truth values of the variables in it. I point this out because if "and" and "or" are chosen to satisfy the and/or distributive property, then Elkan's identity is exact in both crisp and fuzzy logic.[12]

>So, if the and/or distributive laws hold,
Yeah, they hold for fuzzy sets.
>then the equivalence holds.
No. Only if the Excluded Middle law holds as well, which it doesn't. That's the whole point behind fuzzy logic in the first place. It would appear that this "proof" only shows that fuzzy logic collapses to classical logic because it's classical logic in its assumptions, not fuzzy.[13]

Elkan demonstrates that every consistent fuzzy system has exactly two truth values.[14]

> t(A or not A) = t(true)
by the last axiom
∧∧∧∧∧∧∧∧∧∧∧∧∧∧∧∧∧∧∧∧∧∧[15]

. . . it looks to me like your problem is in the underlined section. That formula is only correct if t(A) = 1, which is far from a given. . . . All I see here is that if you put binary values into fuzzy logic, you get binary values back out again. This is not news.[16]

The interventions cited above express radically opposed viewpoints on Elkan's theorem and his demonstration. While some deem the theorem perfectly correct, others declare that it is in no way correct and put forward contradictory reasons in support of this assertion: an error made in the introduction of the principle of distributivity of "and" with respect to "or," an error linked to the implicit introduction of the principle of the excluded middle, a mobilization of inadequate hypotheses, the irrelevance of the notion of logical equivalence used, recourse to incorrect numerical equalities, the illegitimate importation of binary values into fuzzy logic . . . Elkan's result enters from the outset into what De Millo, Lipton, and

---

[12] L. Petrich, August 2, 1993, cited in message 1823.

[13] S. Kroger, August 2, 1993, cited in message 1823.

[14] M. Ginsberg, October 25, 1993, cited in message 1405.

[15] It seems clear that the repetition of the symbol "∧" is intended simply to stress the expression placed immediately above.

[16] D. F. Hefferman, message 834.

Perlis call, rightly or wrongly, the great majority of theorems that are contradicted, disallowed, thrown into doubt, or ignored.[17]

We must note that, over and beyond their disagreements, the participants in the debates seem to share a common vocabulary and a mode of questioning whose emergence we observed in teaching situations. In particular, we again find references to logical principles such as the principle of the excluded middle and the principle of distributivity, to operators of conjunction and disjunction ("and," "or"), and to notions of equivalence, binary logic, and so on. This agreement on the terms of the dialogue induces a certain complicity in the conduct of the debates. The situation is hardly surprising, if we recall that the participants are primarily either North American academics affiliated with computer science departments or researchers in computer science working for private enterprises in the United States.

However, once again we also encounter a great mix of reactions to a demonstration mobilizing a formalism, quite like the confusion we saw among apprentice logicians. In fact, the formal presentation of Elkan's proof does not appear sufficient to produce a consensus as to the correctness of the result. It does not succeed, either, in bringing about a consensus as to the nature of the "error" its author may have committed. To the extent that the formalism provided does not suffice as such to produce agreement among the parties, the logical proof is disputed—as surprising as that may seem to anyone who associates "logical" with "indisputable." This situation warrants closer analysis.

The debate triggered among the protagonists appears intimately connected to the existence of antagonistic visions of what Elkan's proof mobilizes, but also to antagonistic visions of what fuzzy logic must be. The controversy bears simultaneously on the logical notions and principles that Elkan has chosen to use in his demonstration and on the way they are displayed. We shall return to the question of display later on. As for the question of choice, the actors put forward the logical principles that fuzzy logic must in their view accept or reject (in particular the principles of distributivity and of the excluded middle). The divergence of views on these issues represents a first dimension of the debates. The writers' differences constitute and prefigure both antagonistic performative definitions of fuzzy logic and also representations and affiliations that are more or less distant with respect to the work of the most prominent researchers in the field.

Thus it appears at the outset that fuzzy logic is not the object of any central, consensual definition. As the expression "fuzzy logic" is used to characterize perceptibly distinct objects, depending on who is intervening,

[17] De.Millo, Lipton, and Perlis, "Social Processes," 272.

we may be tempted to grasp it in the first place as an emblem behind which actors who endow it with variable significations can rally. The expression would be better described, however, using the terms of the historian Alain Boureau, as a collective statement. This conceptual tool is helpful here as a way of accounting for the collective use of a single expression that links objects, and coordinates the actions of a set of actors in a complex way; although the expression is appropriated in a distinct fashion by each individual, its common usage induces specific relational dynamics.[18]

The plurality of formal systems associated with the expression "fuzzy logic" does not depend solely on the competing claims of exclusive definitions of this object, however, even if such claims predominate. Certain participants in the debates stress in a performative way the multiplicity of the axiomatics that can be legitimately associated with fuzzy logic, according to them, or, more accurately, with fuzzy logics, as soon as these logics authorize an infinite number of truth values. Thus one writer declares that certain fuzzy logics accept the principle of the excluded middle, contrary to Elkan's assertion:

> The author claims, in the discussion following his theorem, that "[w]hat all formal fuzzy logics have in common is that they reject at least one classical tautology, namely, the law of the excluded middle." . . . Even such a claim about the encompassing lack of validity of the law of the excluded middle is false as can be seen by considering Lukasiewicz's continuous logic L-Aleph-1, which satisfies both the laws of the excluded middle and contradiction while failing to satisfy idempotence[19] and distributivity properties.[20]

By affirming that a logic attributed to the logician Lukasiewicz[21] is a legitimate form of fuzzy logic, the author of this message is negotiating,

---

[18] See Alain Boureau, "Propositions pour une histoire restreinte des mentalités," *Annales ESC* 6 (1989): 1491–1504. See also the way the notion of collective statement is used to account for the construction and the role of the expression "scientific policy" in France immediately after World War II, in François Jacq, "Pratiques scientifiques, formes d'organisation et représentations politiques de la science dans la France d'après-guerre: La 'politique de la science' comme énoncé collectif (1944–1962)," Ph.D. diss., École Nationale Supérieure des Mines de Paris, May 1996.

[19] The property of idempotence generally refers to the identity between a proposition "P" and "P or P" (idempotence for the disjunction) or between "P" and "P and P" (idempotence for the conjunction). Let us note that so-called "linear logic" calls this postulate into question. Its axiomatics is inscribed, as with any logic, in a set of specific practices (as Wittgenstein's theses emphasize). The rejection of this identity is inscribed especially in a practice of developing software for the automatic demonstration of proofs, in which, schematically, the operation that consists in writing "P" is not the same as the one that consists in writing "P and P." See J. Y. Girard, "Linear Logic," *Theoretical Computer Science* 50 (1987): 1–102.

[20] E. Ruspini, message 816.

[21] See the work of this logician on multivalent logics. While the expression "multivalent logic" has already been introduced, it is important to specify that the qualifier "continuous"

properly speaking, the definition of a fuzzy logic (or logics). Such negotiations are commonplace in the research world, in contrast to the schoolboy expectation of a definitive judgment on a proof, and of stable, univocal, and universally recognized definitions of logical notions and theories such as fuzzy logic. The intervention of a student who is inquiring in a polite and respectful tone about the outcome of the controversy over Elkan's proof provides a good illustration of this phenomenon:

> Dear sir/madam:
> I am a Ph.D student in Department of Systems at Binghamton University (SUNY at Binghamton). I am very interested in the discussion about Elkan's paper. I want to know the current status of the discussion. Is there anyone who can help me? Thank you very much.[22]

In the same way, here is how one actor denounces the existence among philosophers of a sort of stockpile of wares, and calls for a definitive viewpoint on the meaning to be attributed to the notion of logical equivalence:

> So, what exactly is meant by "equivalent"? This question should not be left for philosophers to endlessly argue about but should be made clear once and for all.[23]

The examination of these interventions brings to light behaviors and expectations (in particular, the expectation of univocity) that constitute the background for the teaching practices we examined earlier.[24] These behaviors and expectations are an extension of modes of interaction characteristic of the dynamics of learning that we have analyzed, on the same basis as the use of normative definitions and a common (or at least widely shared) vocabulary and mode of questioning. We now need to look more closely at certain salient features of the questions formulated by the actors.

The Heterogeneity of Ways of Doing Logic

When we analyze the messages devoted to Elkan's theorem, we note that disputing the choice of logical principles is a perfectly legitimate undertaking in the eyes of virtually all participants. The writers agree on the whole

---

refers here to an infinite set of truth values forming a continuous spectrum. See Jan Lukasiewicz, *Selected Works*, ed. Ludvik Borkowski (Amsterdam: North Holland, 1970).

[22] B. Yan, message 1383.

[23] H. Lucke, message 827. This passage has to be added to a list, just beginning to take shape here, of denunciations of "bad" practices and representations of logic that are said to be conveyed by philosophers.

[24] Lakatos would probably see these as marks of authoritarianism in the teaching of science. See *Proofs and Refutations*, p. 143.

not to deem any given logical principle necessary in itself. For most of them, the question of the "relevance" and the "correctness" of the axiomatics explicitly and implicitly mobilized in Elkan's proof constitutes a stake in the debate that cannot be called into question.

This discussion consists in an advancing of resources that the protagonists consider to be associated with the text of the proof (notions, logical principles, and so on). I use the expression "advancing" to characterize the work that consists in bringing the foregoing resources and associations to light and representing them as essential. Such is the case, for example, when a participant seeks to emphasize the implicit recourse to logical principles in Elkan's proof.

Let us note here again that such work is neither "nonmaterial" nor situated solely in the minds of the participants. It consists first of all in the production of a series of inscriptions on a computer keyboard, followed by their transmission to an electronic forum. This production of inscriptions is in fact constitutive of the controversy under study.

This being said, we may observe that the use of a formalism, far from making all the resources of the proof, and their coherence, "explicit" (and thus far from bringing immediate clarity to Elkan's demonstration), helps condense associations that are "implicit" for authors and readers alike. It puts a great deal of tacit knowledge into play simultaneously,[25] which explains why the exchanges are so complex. If we reformat the notion of "informal thought" as Lakatos used it, to qualify the explicitation and negotiation of tacit practices and "implicit" knowledge, then Lakatos's thesis according to which informal thought circumvents formal thought is quite applicable here.[26]

In the case that interests us, we are nevertheless obliged to note that, for a number of actors, the field of negotiations is relatively limited. Examination of the messages shows that a specific, targeted kind of doubt lies behind the exchanges. This type of doubt has much less to do with a logicist tradition,[27] which would lead in particular to advocating

---

[25] On the notion of tacit knowledge, as a capacity to use "nonexplicit" skills, see Michael Polanyi, *Personal Knowledge: Towards a Post-Critical Philosophy* (London: Routledge, 1958), and Polanyi, *The Tacit Dimension* (New York: Anchor, 1957).

[26] Lakatos, *Proofs and Refutations*.

[27] The term "logicism" usually refers to a program attributed to Frege and Russell for founding a unitary, totalizing logic that would underlie all "reality." It is opposed to a way of constructing "logic" in the form of distinct models with differing relations and domains. This notion, to which historians of logic attribute varying meanings, would warrant a nuanced expression in the context of a fine-grained construction of the history of logic. It is used here chiefly in this specific sense, inasmuch as it makes it possible to emphasize two distinct approaches that have precedents and that constitute elements of historically situated debates. Concerning the construction of this "grand" logic, see especially Gottlob Frege, *Translations from the Philosophical Writings of Gottlob Frege*, ed. Peter Geach and Max

the intrinsic necessity of one particular logical principle or another, than with a formalist tradition, which would incite in particular a focus on the question of the overall coherence of the set of principles mobilized.[28] Here we are very far from the arguments of a philosopher of logic such as Husserl, who sought to "found" logic on the universal necessity of logical principles such as the principle of contradiction (which is a close relative of the principle of the excluded middle). Here, many protagonists abandon this principle without even feeling a need to offer justifications.[29] The participants in the Elkan debate use definitions and practices of logic that are quite distant from the definitions and practices that some sociologists of knowledge, philosophers, or psychologists attribute to logic, while granting a central, stable place to a certain number of syllogisms or logical principles.[30]

Moreover, the fact that the type of targeted doubt that animates this debate is much closer to formalist than to logicist questioning (in the

---

Black (Oxford: Blackwell, 1959). See also, in the face of the so-called paradoxes of self-reference encountered by "grand" logic, the construction of a theory of types in pursuit of this effort in Bertrand Russell and Alfred North Whitehead, *Principia Mathematica* (Cambridge: Cambridge University Press, 1910).

[28] There is no consensus among historians and philosophers of logic about the meanings to be attributed to the notion of formalism (or to the notion of logicism). The term "formalism" is often used, however, with reference to a program attributed to Hilbert, and with reference to the construction of a logic that rests on the definition of systems of objects through the specification of their relations. On this subject, see especially David Hilbert and Wilhelm Ackermann, *Grundzüge der theoretischen Logik* (Berlin: Springer, 1928). At this stage, the latter notion of formalism is used here, as opposed to the notion of logicism, in this specific sense. Later, we shall return to this notion and attach other significations to it, in order to characterize specific skills and practices of writing, in an extension of the analysis of the teaching of logic. Thus the reader will need to be especially careful not to invest this historically charged term with meanings that are foreign to it (the same thing holds true, moreover, for the notion of "logicism").

[29] See especially how Husserl describes the self-evidence of the principle of contradiction starting from the lived experience of the meaning of the notion of contradiction, in *Logical Investigations* , pp. 46–47, 63–64.

[30] This distance cannot be grasped without taking more generally into account the considerable gap that separates the logical skills of researchers in the humanities and the social sciences from those of researchers in logic, a gap that seems only to have deepened since the late nineteenth century. Taking such a gap into account constitutes a key to reading the continuity that can be discerned between various problematics in the humanities having to do with logic, problematics developed throughout the twentieth century on the basis of relatively "stagnant" representations of logical objects and a perceptible lack of information about the evolution of the debates in logic since the end of the previous century. On this subject, see once again Lévy-Bruhl, *Primitive Mentality*, and Hamill, *Ethno-logic*. Under these conditions, if the present study makes it possible to renew certain representations of logical objects and to bring elements of reflection to the topic, especially for sociologists who think in terms of rules or reasoning "imposed" by logic, it will not have been in vain. We shall come back to this point.

senses defined above) underlines the thesis according to which the participants in this controversy have benefited from a certain stability in the way
logic has been taught in the United States since the end of World War II.[31]
Let us recall that we are dealing for the most part with researchers working in computer science in universities or private companies located in
the United States. This fact is fundamental in accounting for the relative
homogeneity of the tacit knowledge that allowed the debate to unfold.

To adopt an analytic posture comparable to Warwick's,[32] among others, we may note a strong convergence of tacit practices, the origin of
which can be sought in the rather broad diffusion of a particular way
of teaching.[33] The high degree of this convergence, combined with the
incomparable means for intervention and interaction offered to the various actors by the electronic newsgroup,[34] allows a very broad multilogue
among the participants. It also means that we can escape here from the
scenario represented in Davis and Hersh's descriptions as a very narrow,
"corporatist" structuring of exchanges.[35]

In order to emphasize the broad consensus that is instituted, given the
mobilization of this targeted doubt, it may be useful to look closely at the
expression of a marginal position and to analyze the reactions it triggers.
Confronting the entire set of views expressed in the electronic forum so
far, a certain H. Lucke intervenes to protest the way various interlocutors
"attribute meaning" to the abandonment of a formulation of the principle
of the excluded middle.

> This doesn't make sense to me. I cannot imagine a scenario in which I was
> uncertain about the truth of (B or not(B)).[36]

Ater declaring that he "does not understand" the meaning of this abandonment, the author of the message asserts that he "does not see" how
abandoning the principle of the excluded middle can be of use in the
practice of computer science for managing a database. He then develops
a scenario to show that this approach "has no meaning":

> Nor can I see where the advantage of a reasoning system that allows t(B or
> not(B)) < 1 lies. If my database contains, among others, rules like if B then
> perform action A and if not(B) then perform action A then I would like to

---

[31] See Rosental, "De la logique au pathologique."

[32] Warwick, "Cambridge Mathematics and Cavendish Physics," part 1.

[33] One must agree with Pierre Bourdieu that "the sociology of education is a chapter, and
not a minor one at that, in the sociology of knowledge." Bourdieu, *The State Nobility: Elite
Schools in the Field of Power*, trans. Lauretta C. Clough (Stanford, Calif.: Stanford University Press, 1996), p. 5.

[34] This point will be developed more fully later on.

[35] Davis and Hersh, "Rhetoric and Mathematics."

[36] H. Lucke, message 813.

perform A independently of my current truth value of B. Or to give you a real life example: Suppose you are making a bet with a friend about the outcome of a soccer game, and you agree to settle the bet immediately after the end of the game. That is if you win your bet you have to go to your friend to collect the money, but if you lose, you have to go to him to pay him your money. In either case you have to arrange to see your friend and you know this even before the end of the game. It would be stupid to say: "I don't know the outcome of the game yet, so I don't know whether I have to see my friend."[37]

Whatever the reasons (and they may well have to do in part with the specificity of his academic itinerary), Lucke does not fall into step here with the consensual position of his interlocutors; instead, he denounces the arbitrary and absurd character of that position, developing his own argument along two different lines. The first consists in affirming that giving up the principle of the excluded middle is in theory "unthinkable."[38] The second consists in putting a little story on display to convince his readers of the impossibility of giving up this principle.[39]

In other words, in order to show that giving up the principle of the excluded middle is unthinkable and impossible, like other philosophers of logic before him[40] this writer is not content to invoke self-evidence, self-sufficiency of a "purely logical" type. He ultimately mobilizes different demonstrative resources, basing the "necessity" of the principle in question on a story staging a practice that is thereby seen to be associated with the logical principle. This approach, which could again be characterized as "logicist," thus rests here on a specific dynamics: the inescapable self-evidence of the principle of the excluded middle is affirmed even as this principle is being justified, by means of a scenario deemed exemplary, the scenario being offered as an ultimate resource for convincing the forum's participants and even for forestalling a possible accusation of dogmatism.[41]

---

[37] Ibid.

[38] This approach must be compared with the one Edmund Husserl uses extensively (Husserl, *Logical Investigations*).

[39] Let us note the parallel between this approach and the one adopted in Plato's *Dialogues*.

[40] For Schopenhauer, it was as impossible to think against the principles of identity, contradiction, excluded middle, and sufficient reason, as—to use his own image—to move our limbs in a direction opposed to their articulations. See Arthur Schopenhauer, *The World as Will and Idea*, trans. R. B. Haldane and J. Kemp (London: Trübner, 1883–86 [1819]).

[41] We should note that this strategy does not always meet with success; see Rosental, "De la logique au pathologique," for many more cases in which such modes of justification fail, leading to accusations of arbitrariness analogous to the one analyzed above in a teaching situation starting with the formalization of the expression "the current president."

By identifying two quite specific demonstrative practices and situating them in a continuum, we note, just as Eric Livingston did,[42] the need to take into account a splintering of demonstrative practices. Correlatively, a vocabulary is needed to characterize these practices: hence the recourse to the expression "appeal to self-evidence" and the reference to the constitution of "scenarios."

The use of the term "scenario" highlights the material dimension of a textual production. Lucke in fact is putting on display a *form of life* (a situation involving a "bet among friends," or the management of a computer database) in order to establish the necessity for the principle of the excluded middle.[43] This form of life is mediated by a textual scenario according to a procedure found in Plato's *Dialogues*, as well as in the methods for teaching that we have observed.

This is not to say that the logical operation associated with a manipulation of objects in the context of the scenario retained becomes its "reflection," with no impact on the way this manipulation and the objects in play are grasped. Rather, in this associative operation a co-construction takes place: the representation of the logical operation and that of the manipulation of objects put forward in the context of the scenario are transformed simultaneously.[44] Both the management of a computer database and the principle of the excluded middle are redefined or transformed by the association made between these elements in Lucke's text. The specific way the representation of the naturalizing manipulation is transformed could be characterized as "transformalization," to highlight the effect induced by putting forward a link between this manipulation and a formal symbolic operation.[45]

By making such a procedure of naturalization a resource of his analysis, as we have seen above, Bloor finally puts himself on the same level as

[42] Livingston, *Ethnomethodological Foundations of Mathematics.*

[43] The term "form of life" is used here approximately in Ludwig Wittgenstein's sense, although I adapt it to a certain extent as we shall see shortly. Wittgenstein visibly grasped the way mathematical operations can be justified in practice through the exhibition of *forms of life* that naturalize them. Thus he describes how taking the root of the number "minus one" can become an acceptable operation after a translation in terms of vector manipulation is associated with it. See Ludwig Wittgenstein, *Wittgenstein's Lectures on the Foundations of the Mathematics*, ed. C. Diamond (Hassocks: Harvester, 1976), p. 226.

[44] On this subject, see also Michael Lynch, "The Externalized Retina: Selection and Mathematization in the Visual Documentation of Objects in the Life Sciences," *Human Studies* 11 (1998): 201–34; Lynch, "Discipline and the Material Form of Images: An Analysis of Scientific Visibility," *Social Studies of Science* 15 (1985): 37–66; Michael Lynch and Steve Woolgar, eds., *Representation in Scientific Practice* (Cambridge: MIT Press, 1990); Bruno Latour, "Visualisation and Cognition: Seeing with Eyes and Hands," *Knowledge and Society* 6 (1986): 1–40.

[45] My own use of the expression "form of life," evoked above (note 43), includes this notion of "transformalization."

the actors whose behavior he wants to study.[46] He becomes a potential interlocutor, both of these actors and of certain philosophers, proposing an object construction to compete with theirs. But in so doing he does not make it clear how such an approach can constitute an object of analysis. In particular, he does not offer the possibility of understanding how the naturalizing associations or identifications of the actors enter into the constitution and legitimization of symbolic manipulations that stem from specific practices (for example, original practices used in the development of computer software).

From a methodological standpoint, as we have already noted, this must incite us to give up introducing naturalizing operations artificially into our narrative in order to "explain" the actors' logic. Such an introduction would in effect mean transforming or transposing the actors' objects. It would prevent us from grasping the underlying principle and the modalities of an approach that is in fact characteristic of the actors, an approach that proves essential to the elaboration of their objects.

But let us return to the analysis of the last excerpt cited. We may note that, beyond his singular attachment to the principle of the excluded middle, the writer adopts a procedure for putting forward a logical necessity that was abundantly used in the methods of teaching logic we examined above. The sedimented character of this tacit practice can also be observed in the message of the actors who propose a counterdemonstration of the result advanced by Elkan. The routine nature of this type of quasi-theatrical staging can thus be brought to light.[47]

By way of response to the preceding message, a text by Y. Tanaka mobilizes for its part, too, a scenario staging a form of life, this time in order to show, to the contrary, the need to abandon the principle of the excluded middle in certain "situations." Precisely to the extent that he displays the need for such renunciation in the context of specific practices, this writer, unlike Lucke, does not adopt the position that I have characterized as "logicist." Tanaka is in fact trying to show that giving up the principle of the excluded middle creates essential maneuvering room in the elaboration of a fuzzy expert system, thus making it possible to deal with cases of conflict between rules.[48] To this end, he counters Lucke's scenario with another one about the way a color photocopier works.

---

[46] Bloor, *Knowledge and Social Imagery*.

[47] See also the works of Simon Schaffer on proofs put on stage: "Universities, Instrument Shops"; and "Self-Evidence," *Critical Inquiry* 18 (Winter 1992): 327–62. The reader may also want to consult Erving Goffman's analyses of the staging of everyday life: *The Presentation of Self in Everyday Life* (Garden City, N.Y.: Doubleday, 1959).

[48] We shall not attempt here to open the black boxes put into play in these descriptions of expert systems. That task would require importing a great deal of data, an effort that is not justified in the context of the present problematics.

As in the earlier case, the "necessity" of abandoning the principle of the excluded middle is put forward through the textual exhibition of a singular practice (itself transformed or "transformalized" by the association the author makes between it and a logical operation). This practice lies in the automatic adoption by a photocopier of limited contrast for a zone of an original that is neither entirely black nor entirely "nonblack" (i.e., gray), and strong contrast for a zone of an original that is entirely black or entirely "nonblack" (i.e., white). The author suggests that the "abandonment" of the principle of the excluded middle in fuzzy logic makes it possible to use rules that in turn make such a way of working possible:[49]

> You're still thinking Boolean. . . . This seems like a Boolean situation. Before the game, it's probabilistic; after the game, it's 1 or 0. May I suggest another example. You have some kind of contrast adjustment on a copier (this is fuzzy). Say you wanted the contrast to be high (or low . . . whatever) if the color was black or white, but the contrast level should be different if gray. If (color is BLK) or (color is NOT BLK) --> this rule would yield something close to 1.0 if the color was black or white and something close to 0.5 (depending on how the MF's [membership functions] are shaped) if the color was gray. This is what I might want. With fuzzy logic (if B or not B) has much more meaning than in Boolean (it's meaningless in Boolean).[50]

Like the preceding passage, this one allows us to note to what extent a text as short as the principle of the excluded middle, whether it is formulated in prose or put in symbolic form, must be accompanied by a set of demonstrative resources in order to appear necessary or to be found deficient, or even simply to be "interpreted" (i.e., associated with one or several forms of life). The formulation of such an observation gives a particular meaning to Wittgenstein's thesis according to which the necessity of a rule would be only the "faithful reflection" of the means deployed to "represent reality";[51] or to the argument formulated in similar terms by Jacques Bouveresse: "The necessity is not imposed on us by the nature of the things to which our systems of representation have or ought to have conformed, but solely by the way we have chosen the systems in question."[52] This analysis also allows us to account for a thesis formulated

---

[49] Such at least is the paraphrase that I am constructing in order to "reveal" a "form" (nonsymbolic, this time) of the argument, which some may find difficult to grasp. By carrying out an act of reflexivity in this way, I want to insist once again on the specificity of the material procedures of formalization adopted by the protagonists in the debates.

[50] Y. Tanaka, message 814.

[51] See also the analyses developed on this topic in Alain Coulon, *Ethnométhodologie et éducation* (Paris: Presses Universitaires de France, 1993), pp. 193–228.

[52] Jacques Bouveresse, *La force de la règle* (Paris: Minuit, 1987), p. 15.

by certain logicians according to which there is in fact no such thing as the "neutrality" of a logic with respect to the operations that it attempts to "represent."[53]

Moreover, it is important to stress the fact that, concerning this passage, the appeal to the self-evidence of a logical approach is once again accompanied by the deployment of a scenario deemed exemplary. The approach that consists in advancing the self-sufficiency or the ideal necessity of a logical system is again adopted only with the support of such a resource. Let us recall that we had noticed, in analyzing Lucke's message, how the latter had taken recourse to a scenario deemed exemplary in order to support the assertion of the indisputable self-evidence of the principle of the excluded middle. We see now that this procedure in no way constitutes a guarantee that readers will be won over, that they will accredit the thesis being defended.

In fact, far from adopting the same viewpoint in his message, Tanaka tries on the contrary to relativize the validity of the principle of the excluded middle. To this end he no longer presents the scenario advanced by Lucke as exemplary, as illustrating the indisputable self-evidence of that logical principle; he presents it rather as a simple example of a particular type of reasoning. The approach that consists in making Lucke's scenario shift from exemplary status to essentially exemplifying status, that is, in making Lucke's text appear above all as the production of an example, is at the heart of the deployment of a relativizing point of view: Tanaka denounces a "Boolean attitude" by that very token, and suggests that the validity of the principle of the excluded middle is relative to a type of attitude, to a way of "thinking," that has a particular place in the context of the great fractures that divide logicians, of which Boolean logic represents only one component.

We must note, however, that Tanaka's relativizing attitude toward the validity of the principle of the excluded middle does not extend to all the statements in the debate, particularly his own. Since he adopts a realist's viewpoint toward his own statements, Tanaka—like most of the other participants in the debate, moreover—is not a relativist in the strict sense. As we have seen, the great majority of the protagonists displayed contradictory viewpoints in a tone expressing certainty; the observation of such radical divergences did not lead them to doubt their own positions and in particular did not lead them to thematize doubt by adopting a relativist position.

---

[53] On this topic, see especially Largeault, *La logique*, pp. 24 and 117. See also Willard Van Orman Quine, "Truth by Convention," in Quine, *The Ways of Paradox and Other Essays* (New York: Random House, 1966 [1936]), pp. 70–99.

5everal sociological studies that have focused on debates in the experi-
entaI sciences have brought analogous results to light.[54] This research
us shows that logic is not exempt from the adoption of relativist posi-
ons. Still, one fact is particularly striking in our case: the possibility of
observing two radically opposed attitudes (the one idealizing and the
other relativizing) toward a statement as famous and celebrated (at least
in other times and places) as the principle of the excluded middle. With
only one exception,[55] the participants do not attribute a central and indisputable character to this principle. At the very most, they grant it a place
in the framework of an entirely specific practice, that of Boolean logic.

Correlatively, we are taking the measure of the deep dissensions within
the group of participants in the debate over the way logic should be practiced.[56] If the analysis of the two preceding messages has not sufficed to
illustrate this, the excerpt that follows may be helpful. Its author adopts
a demonstrative procedure identical to Lucke's, but this time in order to
defend an opposing viewpoint that we can characterize as "profuzzy"
idealizing:

> If we know only a tiny part of a system, we can hardly claim that everything
> else, including the major part we don't know, includes into systems comple-
> ment![57] Thus A and not(A) is LESS or EQUAL to 1. . . . My opinion is that B
> or B = B whatever B is, and (assuming the coin is allowed to stay on its corner)
> B or not(B) ≤ 1.[58]

With the same demonstrative resources as those used by Lucke (appeal
to self-evidence, marked here by the use of an exclamation point; recourse

[54] See especially Latour, *Science in Action*; and Latour, *Petite réflexion sur le culte mo-derne des dieux Faitiches* (Le Plessis-Robinson: Synthélabo, 1996).

[55] The exception is the message by Lucke cited above.

[56] My perspective is not denunciatory here but simply descriptive, for while the existence of dissension may appear surprising to certain readers, it will not surprise most logicians; see especially Largeault, *La logique*, pp. 21, 23, 110, 115, 118, and 119. In 1937, Rudolph Carnap had already highlighted major disagreements among logicians as to how logic should be practiced; in formulating his "principle of tolerance" he noted the way logicians tended to reject rival practices by characterizing them as "alogical." The principle of toler-ance aimed at getting apparently contradictory practices, advanced by different logicians, accepted as logical; Carnap sought to show that these practices corresponded to distinct "forms of language" whose syntax had to be specified. He was pleading in this way for the syntactic approach to logic to which he was strongly attached; see Carnap, *Logical Syntax of Language*, pp. 51–52. For an introduction to the approaches to logic that were for a time characterized as nonclassical, when they were not simply *disqualified*, see Jon Barwise and Solomon Feferman, eds., *Model-Theoretic Logics* (New York: Springer, 1985).

[57] The author seems to be saying that "we can hardly claim that everything else . . . is included in the complementary subset of the given set or system."

[58] J. Uski, message 1839.

to a scenario deemed exemplary), the author of this message develops a radically opposed point of view: he asserts that he cannot conceive of taking recourse to a form of the principle of the excluded middle. He nevertheless feels the need to justify the necessity of this abandonment by developing a specific scenario, that of a coin that, after it is tossed, lands on neither of its two sides but rather on its edge (which he calls a corner).[59] Although the two authors mobilize identical demonstrative strategies, their disagreement is obviously complete.[60]

But let us now return to the general case, the one that consecrates the options I have been calling "formalist." If the actors put forward competing axiomatics, as we have noted, just how can they subject them to debate? To answer this question, let us begin by analyzing what criteria of judgment the participants deem it legitimate to mobilize in order to characterize the choices among logical principles and assess their relevance.

The Use Values of Demonstrations

In examining the set of messages devoted to Elkan's theorem, we have noted that the contributors regularly make judgments as to the value of the proofs. This practice is difficult to perceive on a first reading. It sometimes stems from the use of expressions or details that, taken in isolation, might easily be missed. In contrast, once the corpus is taken as a whole, the telling details can no longer go unnoticed.

Thus for example the authors of one message declare that the introduction of certain properties of Boolean algebra into fuzzy logic is "of no interest" and leads to a "trivial" structure.[61] In another message, the writer asserts that it is not only "incorrect" but also "trivial" to introduce certain hypotheses, certain classical tautologies, certain notions of logical equivalence, into fuzzy logic, so that Elkan's proof is at once "too heavy" and "useless":

[59] Such at least is my own interpretation of the scenario evoked in parentheses, for the author does not bring any supplementary clarification on this point in the remainder of his text. The message nevertheless seems clear enough: one can no more demand a response to the question "true or false?" than to the question "heads or tails?"

[60] We must note that the need to abandon the principle of the excluded middle has been posited by various individuals and groups over a very long period of time. For example, the partisans of intuitionist logic do not assimilate the true to the noncontradictory, and they use scenarios that highlight the distinction to be made between arguments about finite sets (attributed to Aristotelian logic) and arguments about infinite sets, in which the principle of the excluded middle "no longer holds." On this subject, see especially Jean Largeault, *L'intuitionnisme* (Paris: Presses Universitaires de France, 1992).

[61] D. Dubois and H. Prade, message 1035.

His theorem is based on incorrect assumptions . . . it is well known that many tautologies of propositional logic fail to be valid in fuzzy logic. Assuming therefore that "equivalent" (in Gaines' Axiom 4) means equivalence in the sense of classical logic is bound to lead to error. . . . One does not need to resort to Elkan's lengthy proof to see that if all tautologies in the classical propositional calculus extended to fuzzy logic, then the only possible truth values are zero and one. . . . each proponent of a formal approach must define logical equivalence—on the basis of the semantics of that formal system—in order to provide meaning to Axiom 4. Elkan, however, puts the cart before the horse, with disastrous consequences.[62]

The author of this message brings to light the existence of sedimented logical practices in order to pass a negative judgment on Elkan's proof. Certain results are "well known," certain ways of practicing logic "must" be respected. The negotiation of ways of practicing logic and the simultaneous devalorization—in the strict sense of the term—of Elkan's demonstration are carried out in a normative mode.

Elkan and some of his supporters seem for their part to acknowledge the terms of the debate initiated by these critics, along with the assertions that touch on the legitimate modes of the exercise of logic. In fact, they reject the charges of incorrectness and triviality leveled against the notion of logical equivalence and counter by stressing its use value. For them, the notion of logical equivalence mobilized in Elkan's proof, which they attribute to Boolean logic, is the most universally used notion in logic. By this token, a logical system that did not use it would be of limited value.

Moreover, certain writers condemn the fact that this notion of logical equivalence is used in practice by researchers in fuzzy logic themselves or, when this is not the case, they object that the latter fail to specify their recourse to a different notion of equivalence. Thus, for example, one contributor writes:

Elkan admits that the fourth postulate (about logical equivalence) that makes this proof possible is controversial, but everyone (including Kosko) uses it.[63]

And another:

If some other interpretation is to be used, this interpretation should have been supplied by Gaines, for boolean logic is by far the most widely used logic system. Any argument to the affect [sic] that boolean logic was not intended but rather some other unspecified (possibly tertiary) logic system sounds terribly like weaseling to me.[64]

[62] E. Ruspini, message 816.
[63] F. Brill, message 788.
[64] H. Lucke, message 827.

These authors are thus trying to justify the relevance of the axiomatic choices on which Elkan's demonstration rests by highlighting the universal character of certain logical practices. The latter would be universal in the sense in which recourse to them would be generalized: researchers in fuzzy logic would be included among their users, Bart Kosko[65] and Brian Gaines[66] being examples. For these authors, in fact, the point is not to stress some ideal universality of results that would undergird these particular logical practices. The point is rather to situate the debate on a terrain where taking into account the universality (in the sense of generalized use) of certain practices is at least as important as an aspiration to ideal universality on the part of the proofs.

By bringing to light widespread tendencies and fractures in the collective exercise of logic, these messages thus make logic appear as a practice to which specific universals correspond, those that are connected with a relatively more extensive use of certain systems with respect to others. We observe that the debate bearing on the choice of logical principles and notions not only puts the correctness of Elkan's proof at issue, but that it also raises a separate question regarding the value of the proof: Is the result "trivial" or not? Is the choice of axiomatics justified by custom?

Contrary to what we might have expected, the question of the correctness of the demonstrations is thus not central in such a debate.[67] Up to now, we have seen that several criteria for judging the value of proofs enter into play simultaneously. These criteria are based on distinct notions: a notion of formal value referring to the question of a proof's correctness, a notion of demonstrative value related to the relevance and scope of the result obtained, and a notion of use value concerning the logical principles mobilized. The corresponding registers and their combinations are of course so many sources of controversy where Elkan's theorem is concerned.

The controversy is not limited to these questions, however. Each participant tends in fact to redefine the very heart of Elkan's proof, of what had been demonstrated, in a back-and-forth exchange between lapidary formulations purporting to capture the essence and mainspring of the result and a critical analysis of the explicit resources of Elkan's proof.

We have begun to see that this controversy does not proceed from oral debate but that it unfolds rather through acts of writing. This point needs

[65] Kosko, "Fuzziness vs. Probability."

[66] Brian R. Gaines, "Precise Past, Fuzzy Future," *International Journal of Man-Machine Studies* 19 (1983): 117–34.

[67] For a discussion of the peripheral character of the notion of truth in logic, in the specific context of the establishment of a theory of signification, see Michael A. E. Dummett, *Truth and Other Enigmas* (Cambridge: Harvard University Press, 1978). See also Largeault, *La logique*, p. 22.

to be stressed here, and further developed through a particular focus on the nature and content of the acts in question. It is essential to perceive that this controversy could not be correctly apprehended if the analysis were limited to a study of the argumentation. As we shall see, an important part of the protagonists' activity consists in organizing and managing the material unfolding of the debates. Thus the mediations that constitute the controversy must be brought to light and analyzed.

Among these mediations, all the material resources[68] of reference, all the technologies of denotation,[69] must be carefully studied if we are to circumscribe the elements that facilitate or constrain the debates. Moreover, we shall have to grasp the way the Elkan controversy not only relies on the use of technologies of denotation but also takes these technologies as objects. If a controversy arises over the choice of logical notions and principles, it does so above all because the actors make these notions and principles appear, put them on display, and designate them by producing inscriptions; these actions then trigger further debate over the way the appearances or displays are orchestrated.

DE-MONSTRATING AND APPEARING

The Practice of Substituting Proofs

In the messages addressed to comp.ai.fuzzy, we note that the participants make abundant efforts to produce new demonstrations—or counterdemonstrations—of the theorem in order to convince their interlocutors of the correctness—or the absence of correctness—of Elkan's proof. Indeed, we observe a veritable inflation in the texts through whose mediation the writers seek to make themselves spokespersons for Elkan's article. The writers present their own messages as reformulations of the original proof, a proof to which they attribute recourse to specific resources (assorted logical principles and notions, implicit axiomatics, demonstrative mechanisms); they offer the forum's readers substitutes for the article Elkan published in the proceedings of the 1993 conference on artificial intelligence.

---

[68] Let us recall that the term "material" is used here above all when there is reason to suppose that the element thus qualified may not spontaneously appear as such in the reader's eyes.

[69] I am referring to all the mediations that contribute to giving meaning to the inscriptions, beginning with their material arrangement. On this point, read in a minimalist materialist perspective, see also Jacques Bouveresse, *Le mythe de l'intériorité: Expérience, signification et langage privé chez Wittgenstein* (Paris: Minuit, 1976), p. 97: "The fact that [a proposition] is composed of elements in such and such a number, relating to each one in a given way, confers on it an internal potentiality for signification, in other words for being affected to the representation of a certain type of situation."

Thus we can apprehend the exchanges about Elkan's theorem as stemming in the first place from a production of translations of the original proof, each translation claiming faithfulness to Elkan's text. Instead of focusing on a single demonstrative text, the debate evolves toward a scattering of viewpoints around a growing number of demonstrative texts; the "discussion" bears not only on Elkan's text but also on all the self-proclaimed substitutes for it.

For an illustration of this point, let us examine the message of a participant who is reformulating or translating Elkan's proof. After setting forth what he deems to be the essential aspects of the hypotheses mobilized by the author, he proposes a new demonstration of the theorem:

> I won't bother typing in the rest of the proof, but you can convince yourself of it by checking out a few values for $t(A)$ and $t(B)$ and working out the truth values in the equation. It only works out if you choose values for $t(A)$ and $t(B)$ according to the theorem, otherwise the equation is false.[70]

The author of this message is substituting a new demonstrative text for Elkan's proof. To convince his readers of the correctness of the theorem, he proposes a new demonstration, which both proceeds from an appeal to, and serves as a way to generate, trust.

An appeal to trust, first of all, because the author of this message asks his readers to accept the substitute proof simply because it would be tedious (for the writer and the reader) to retranscribe the original text (and to read it). But also a way to generate trust, based on a declared absence of counterevidence, inasmuch as the writer invites his readers to convince themselves of the correctness of Elkan's theorem not by reading the original proof but by following a very different demonstrative path: the introduction of some values in an equation, as a trial—in other words, the numeric verification of the theorem's correctness on the basis of a few examples.

The examination of this case thus offers an empirical validation of Davis and Hersh's testimony about the role that an appeal to trust and the lack of counterevidence can play in the deductive sciences.[71] This phenomenon of offering substitute proofs, which recurs in many of the other messages, constitutes an essential element in the debates surrounding Elkan's proof. To the extent that this mode of material production of inscriptions organizes the unfolding of the exchanges, we can readily understand why these debates cannot be treated simply as exchanges of oral arguments.

---

[70] F. Brill, message 788.

[71] See also Steven Shapin's analyses of the role that can be played by trust granted to testimony in the recognition of results in the inductive sciences, especially in *A Social History of Truth*.

Another fundamental dimension of the debates can now be brought to light. It is no longer simply a matter of choice but of competing displays of logical principles, whether in Elkan's demonstration or in the new ones produced by the actors.

Making Logical Principles Appear and Disappear in Demonstrations

The substitution of demonstrations based on an appeal to trust (or as a way to generate trust) is not the only mode on which the deployment of new demonstrative texts relies. A controversy that develops over the modalities of display of logical principles in Elkan's proof and its substitutes is itself the source of new demonstrations.

In order to convince their readers that the original proof "presupposes" some particular logical principles, whether to defend the proof or to challenge it,[72] certain writers literally make these principles appear between the lines of Elkan's demonstration or its substitutes. The term "de-monstration," more suggestive than "demonstration," seems appropriate to characterize this specific practice, which consists in producing new inscriptions designed to display something, to make something appear, to show something (a principle at issue in the present case).

For a better grasp of the nature of this practice, let us consider the following example.[73] To de-monstrate that Elkan's proof "implicitly" uses the principle of distributivity, one of the contributors brings this principle into view between the lines of the original text. He produces a series of inscriptions that associate from start to finish an equivalence used in Elkan's demonstration with a formulation of this principle:

> The first statement: not(A and not(B)) = not(A) or not(not(B)) = not(A) or B. The second statement: B or (not(A) and not(B)) = (B or not(A)) and (B or not(B)) = (B or not(A)) and (TRUE) = B or not(A). So, if the and/or distributive law A and (B or C) = (A and B) or (A and C) and A or (B and C) = (A or B) and (A or C) hold, then the equivalence holds.[74]

---

[72] Let us note that Lakatos distinguishes between two ways of challenging a demonstration: a local challenge and a global challenge. At this stage, our analysis situates us in the first of these two cases (Lakatos, *Proofs and Refutations*).

[73] My own methods of demonstration put into play resources comparable to those used by the participants in the debates, except that I do not take recourse to symbolic demonstrations. My own activity implies de-monstrating: bringing to light, giving examples, causing resources mobilized by the actors to appear in order to de-monstrate, and so on.

[74] L. Petrich, message 804.

The author of this message starts with the two members of a logical equivalence mobilized in Elkan's demonstration[75] to exhibit Elkan's recourse to the principle of transitivity. As the two members of this equivalence are respectively "not(A and not(B))" and "B or (not(A) and not(B))," the writer shows that, in order to "end up" with the same expression "B or not(A)," it is necessary to proceed to a symbolic manipulation that is nothing but a formulation of the principle of distributivity. His de-monstration consists in developing the two terms of the equivalence by bringing to light the symbolic manipulation authorized by the principle of distributivity. "Developing" the expressions here means producing new inscriptions and presenting them as equivalent to the first ones, and marking these equivalences with the "equal" sign ("=").

We are reencountering here a practice of "causing to appear" that we have already seen in the context of teaching logic. The "reasoning" put into play, far from being simply localized in the actors' "minds," consists in a deployment of inscriptions and calls for visual attention on the reader's part in order to recognize the "correctness" of the translations carried out. This deployment is neither arbitrary nor the product of pure convention. To be sure, it mobilizes tacit sedimented skills in order to be produced or followed. But it constitutes a concrete support to guide readers, to try to lead them to acquiesce in the translations by visual verification of the equivalences displayed.

What is at work here benefits from being apprehended in the light of the model of distributed cognition developed by Hutchins[76] to pinpoint the role that the use of formalisms[77] plays in the coordination of a group of actors. As we have seen, for Hutchins the manipulation of formalisms constitutes a concrete activity that cannot be grasped simply by following the evolution of mental states. In particular, in his view, the action of the actors' hands and eyes must not be neglected. Moreover, Hutchins does not use the term "cognitive" in opposition to the terms "perceptual" or "motor."

But more fundamentally still, Hutchins's analysis suggests that "thinking" consists in attempting to accomplish tasks that become easier and easier as certain of them are delegated to third parties or technological devices. Such a dynamics of delegation of course implies an important redistribution of skills. In this picture, human beings appear above all as individuals who have recourse to a vast array of mediations in order to act. They appear in particular as good "processors" of symbols—

---

[75] We shall see later on, by looking closely at the details of Elkan's proof, that this equivalence is indeed used in Elkan's article.

[76] Hutchins, *Cognition in the Wild*.

[77] A term whose meaning is close here to the one Latour gives it; see above, p. 50.

and symbols, for their part, represent particularly efficient tools for coordinating action.

Under these conditions, the study of cognition lies for Hutchins in the first place in field observation of the dynamics of collective action and in the analysis of trajectories of representations that are propagated by different mediations. The cognitive research advocated by the author is thus inseparable from research on the "social organization" of the groups considered. It is clear that to the extent that it does not dissociate the analysis of social phenomena from the analysis of cognitive acts, Hutchins's cognitive science cannot be constructed in laboratories, in particular, by carrying out "experiments" on "individuals."

The value of Hutchins's analyses is of course not limited to the use of formalisms by members of a ship's crew (the case studied in *Cognition in the Wild*). His approach may be adopted just as well to study the manipulation of formalisms in the context of a dynamics of production of certified knowledge in logic where scientists are the actors.

To be sure, the system formed on the basis of computers and participants in an electronic forum differs from the configurations studied by Hutchins in the sense that it brings into play a collective dynamics animated by competing projects of orchestration. These latter lead to a limited coordination of the action, as we shall see more clearly later on. To the extent that the actors here are very far from being "collaborators," the description of the way the group works cannot be made simply in terms of coordination; it requires additional resources as well.

The fact remains nevertheless that, in the case that interests us, Hutchins's approach allows us to grasp the way de-monstrative action, which involves the collective of readers and authors participating in the newsgroup, is coordinated by the presence of textual devices; this action is distributed among the inscriptions, the tacit practices of causing to appear, and the acts of visual verification of symbolic manipulations. By this very token, this approach makes it possible to escape from a reductionist localization of the action inside the protagonists' minds. It also offers the means to go beyond an approach stemming from a form of social relativism that would derive from a particular reading of Wittgenstein consecrating the plasticity[78] of texts that put formalisms to work; from this standpoint, such texts would constitute texts "like any other," and one could settle for studying the way "communities" of practitioners work without taking an interest in their symbolic production.[79]

---

[78] I use the term "plasticity" with reference to the thesis according to which logical or mathematical objects would offer no specifically "objective" resistance: it would be possible to make them "bend" at will in order to put forward results that would stem from interests on the part of actors involved in purely human interactions.

[79] See above, pp. 32–33.

But let us come back to the analysis of the previous excerpt from message 804. To the extent that its goal is to bring to light or put on display the mobilization of the principle of transitivity in Elkan's proof, we are indeed dealing with a de-monstration (in the sense in which this term is being used here). Earlier, in a pedagogical context, we saw that one could be taught how to make something appear (logical symbols, logical principles); here, the writer has understood how to make the principle of distributivity appear by carrying out successive translations of an expression used by Elkan. The staging of this "appearance"[80] is constituted by the linking of inscriptions, and it is guided by the search for agreement among the forum's readers as to the legitimacy of the equivalences advanced. In other words, the writer is seeking to win acceptance for the step he is taking in associating one inscription with the other, in order ultimately to emphasize the original author's recourse to the logical principle identified.

The de-monstration is thus presented as a linkage presupposed by the writer (rightly or wrongly) to be self-sufficient, one that does not require complementary developments. Each stage seems to be perceived as transparent: all its elements are purportedly there to be seen, and thus it would convey its own self-evidence. Since at each stage the writer is content precisely to present a display, we can say that this de-monstration is deployed in a series of showings, or "monstrations." This term offers an evocative characterization of a quite specific practice that can be related to a formalist project of making everything visible, a practice that consists in lining up textual fragments that the writer believes he can simply exhibit without further staging.

The term "monstration" underlines the possibility of calling into question a canonical dichotomy between showing (French *montrer*) and demonstrating (*démontrer*). It is not of course a matter of adopting a denunciatory stance here,[81] but rather of developing a methodological viewpoint capable of accounting, insofar as possible, for the actors' ways of working. It is important to note that the de-monstrative practices identified here can be observed in numerous messages. Let us examine, for example, the way a different logical principle, the principle of the excluded middle, is exhibited by two contributors using similar de-monstrative procedures and starting from the same logical equivalence as the one previously identified in Elkan's demonstration. It will be useful to juxtapose excerpts from two messages:

---

[80] Although the objects that appear here may seem very remote, the situation we are analyzing can be compared to the ones described by Élisabeth Claverie concerning the appearances of the Virgin Mary, in *Les guerres de la vierge: Une anthropologie des apparitions* (Paris: Gallimard, 2003). See also Daston, "Marvelous Facts."

[81] Cf. Hardy, "Mathematical Proof."

Elkan needs the law of excluded middle to show the logical equivalence of his implication sentences, e. g. $A \Rightarrow B = (\text{not } A) \vee B = ((\text{not } A) \vee B) \vee 1 ! = ((\text{not } A) \vee B) \vee ((\text{not } B) \vee B) = B \vee ((\text{not } A) \text{ and } (\text{not } B)).$[82]

Elkan's proof—based on the equivalence, in conventional propositional logic, of the formulas "NOT-(a AND NOT-b)" and "b OR (NOT-a and NOT-b)," actually assumes the validity of the law of the excluded middle. (To see this, simply expand "b OR (NOT-a AND NOT-b)" and note the conjunct "b OR NOT-b").[83]

In each of these passages, a different logical principle, that of the ex-cluded middle, is exhibited once again in the de-monstrative trial of a production of inscriptions. Each textual development can be compared to a user's manual.[84] Indeed, although the goal sought here is not to make a device work, a set of procedural instructions is nevertheless provided to the reader with the help of standard logical symbols, corresponding to the execution of an equivalent number of visual verifications, so that the principle of the excluded middle can finally be seen.

In the first message, this principle appears at the heart of a development that almost connects the two terms of the equivalence mobilized by Elkan, "(not A) ∨ B" and "B ∨ ((not A) and (not B))." Almost, for B. Dalton does not take the trouble to display the equivalence between the first mem-ber of Elkan's equivalence, "NOT-(a AND NOT-b)," and the term with which he begins, "A ⇒ B" followed immediately by "(not A) ∨ B." The principle of the excluded middle is thus introduced in two phases. The first phase consists in bringing out the truth value "1," followed by an exclamation point. This punctuation mark is probably intended to em-phasize the fact that the expression resulting from this adjunction is in-deed equivalent to the preceding one, located to its left. But it also un-doubtedly is aimed at stressing, not without irony, the "astute" character of this development. In a second phase, the truth value "1" is replaced by "((not B) ∨ B)." As we have seen, the principle of the excluded middle can be formulated as follows: for every proposition B, (B or not B) is true.

The second message proceeds from a more economical development of inscriptions in order to bring the principle of the excluded middle to light. After displaying the two terms of the equivalence used by Elkan, E. Ruspini suggests that the reader follow directions so as to be able to see Elkan's recourse to the principle of the excluded middle. The text is used once again to lead the reader through the task of visualizing the principle at issue.

---

[82] B. Dalton, August 21, 1993, cited in message 1823.
[83] E. Ruspini, message 816.
[84] On this point, see also the analyses offered in Rotman, *Ad Infinitum*.

In the final analysis, each writer is seeking to provoke an appearance, to get his readers to see what he is trying to show them as the text proceeds. They all use the same method to achieve this goal: the development of an expression in several stages, starting with inscriptions presented as equivalent or equal. This type of text can easily be compared to a user's manual, since it contains a certain number of instructions intended to guide users of the text in their work as readers. The specificity of this user's manual lies especially in the use of symbolic inscriptions that define in part the acts of visual verification to be performed in order to be able to see the equivalences and, finally, the logical principle at issue.

However, we note that the use of the same de-monstrative procedure can lead to different displays. In the last three message excerpts we have analyzed, divergent elements (two different logical principles) presented as essential mainsprings of Elkan's proof have been exhibited. This fact helps reveal the extent to which making the resources of a proof like Elkan's explicit depends in very large measure on the de-monstrative strategies deployed by the authors who re-present it. Making explicit means first of all making apparent.

Moreover, we must note that the expected results do not always emerge from the de-monstrative actions undertaken. The analysis of the messages devoted to these new de-monstrations shows that these latter are apprehended in quite variable ways. While the instructions presented in the de-monstrative texts help define, for readers, the work they need to do, in reality the definitions are no more than partial at best.[85] Not all the protagonists recognize the equivalences put forward as such. They do not all see what the authors are trying to make apparent.

When it is a matter of getting some apparatus to work, the corresponding user's manual often turns out to be inadequate; generally speaking, a large number of tacit practices also have to be mobilized.[86] The same thing is true of the reading tasks assigned to the participants in the debate by the de-monstrative texts. It is clear that the ways in which the writers' instructions are followed vary in relation to the tacit reading practices adopted by the protagonists.

Thus the last message excerpt we analyzed, which constitutes one vector of the appearance of the principle of the excluded middle in Elkan's proof, does not succeed in producing a consensus on what is seen. The participants in the forum do not seem to have received the same training

[85] The readers encountered cannot by this token be identified with the ideal readers described by certain semioticians. On this point, see also the passage from Proust cited at the beginning of part 2.

[86] On the limited usefulness of the instructions, see also Bruno Latour, *Petites leçons de sociologie des sciences* (Paris: La Découverte, 1993), pp. 47–55.

as Ruspini. They do not see what Ruspini, the author of the message, asserts that he sees there himself and is trying to show. To be sure, certain contributors do perceive the appearance that Ruspini is attempting to highlight:

> There was an extensive discussion of this result from Elkan in comp.ai.fuzzy. The consensus (IMHO) was that the proof is inherently flawed due to an implicit assumption of the boolean law of the excluded middle which does not hold in fuzzy logic.[87]

This actor has clearly seen the principle of the excluded middle in Elkan's text, to the point that he does not take into account the divergent views expressed in the course of the debates both on the question of the correctness of the original proof and on the definition(s) of fuzzy logic. Contrary to Lawler's claim, this vision is not in fact the object of any consensus. Certain participants in the debates assert that they "fail to see" the recourse to the principle of the excluded middle, and they go on to produce other displays, as in the following excerpt (where I am using bold type for certain formulas in order to stress the staging of contradictory visions):

> >>Zadeh's logic is distributive and, therefore, the
> >>roots of Elkan's mistake do not lie on
> >>distributivity but on failure of the law of the
> >>excluded middle (as correctly pointed out by
> >>Kroger).
> Well **let's look at this again.** We can simplify Elkan's proof to the following: Starting from Gaines' axioms: $t(\text{true}) = 1$, $t(\text{not } A) = 1 - t(A)$, $t(A \text{ or } B) = \max(t(A), t(B))$, $t(A) = t(B)$ if A and B are logically equivalent, we may consider $t(A \text{ or not } A) = t(\text{true})$ by the last axiom => $\max(t(A), 1 - t(A)) = 1$ => $t(A) = 1$ or $t(A) = 0$. I think the problematic step is the first one, the use of the last axiom. **I don't see the excluded middle coming in here.**[88]

Ruspini's de-monstration has failed for this author, who is trying to show that he has not seen what the previous writer wanted to make him see. By proceeding in this way, he is of course inciting his own readers to adopt the same point of view. We may invoke the politics of the ostrich metaphorically to characterize the attempt of this contributor to align not seeing with not being.[89]

---

[87] D. Lawler, message 853.

[88] H. Lucke, message 827.

[89] See once again the text by Proust cited at the beginning of this part. For an analysis of the major role that witnesses who "have not seen" the essential mainsprings of a demonstration can play in a struggle for the recognition of rival "theories" in the experimental sciences, see Schaffer, "Self-Evidence," and Malcolm Ashmore, "The Theatre of the Blind:

It is useful to compare this type of situation with those analyzed by Thomas Kuhn in *The Structure of Scientific Revolutions*. Let us recall that this historian devoted an important chapter of his book to "revolutions as changes of world-view."[90] According to Kuhn, every change of paradigm brings with it a simultaneous revolution in worldview. Thus Uranus, after having been perceived for nearly a century as a star and then as a comet, was seen starting in 1781 as a planet. For the author, such a transformation in the way an object is seen does not stem from an isolated development. It is not a question, either, of a simple change in the interpretation of a singular phenomenon identified in a stable way. The transformation is produced by an interconnected modification in the way a multiplicity of elements in the world are viewed—that is, by a paradigm shift.

In the case that interests us, the actors are advancing divergent visions of the same proof. The proof in question is not simply the object of differing interpretations. Depending on the logical skills the protagonists have mastered and are using, they see in it, and put on display, different objects.

To account for this situation, there is no need to invoke a possible large-scale paradigm shift.[91] A relationist analysis deployed on a smaller scale, whether or not it adopts the arsenal of general hypotheses associated with Gestalt theory,[92] seems better suited here to interrogate the links that are woven among competing definitions of object systems and divergent ways of seeing.

In fact, the gaps between the different modes of reading a given demonstration can be apprehended as a problem of replication, the difficulty of duplicating the specific mode of reading that the author anticipates. The difficulty stems in particular from the fact that both writer and reader need to have acquired in advance the same large set of specific tacit skills.

The replication of an experimental phenomenon from one laboratory to another usually requires not only similar material arrangements but

---

Starring a Promethean Prankster, a Phoney Phenomenon, a Prism, a Pocket, and a Piece of Wood," *Social Studies of Science* 23 (1993): 67–106.

[90] Thomas Kuhn, *The Structure of Scientific Revolutions* (Chicago: University of Chicago Press, 1962), pp. 110–134. See also Hans Blumenberg, *The Genesis of the Copernican World* (Cambridge: MIT Press, 1987).

[91] This argument could be developed at considerable length. Let us simply note here that, as Canguilhem (following in Bachelard's footsteps) had already observed, the notions of "normal science" and "paradigmatic revolution" are of limited interest for the historian who is confronting a science undergoing continual reconstruction. See Georges Canguilhem, *Le normal et le pathologique* (Paris: Presses Universitaires de France, 1972), as well as the analyses developed in Robert Iliffe, "Theory, Experiment, and Society and French and Anglo-Saxon History of Science," *Revue européenne d'histoire* 2, no. 1 (1995): 73–74.

[92] I am referring to the hypotheses that constitute the point of departure for Kuhn's analysis.

also the organization of a transfer of highly specialized tacit skills among the experimenters.[93] Similarly, the replication of a given type of reading of a de-monstration by a reader necessitates that the reader and writer share numerous and sometimes very unusual skills. In each case, obtaining a replication is more the exception than the rule, given the improbability that such conditions will be met.

Up to now I have been describing the recourse to particular writing procedures and specific acts of visual verification of symbolic manipulations, then the deployment of contradictory visions of the de-monstrations. All these elements converge to emphasize the graphic and visual character of the logical practices with which we are dealing. It is by acts of writing and reading that new objects are produced and set into relation with one another, the latter action leading to a redefinition of the objects re-presented.

We thus reencounter certain characteristics of the work on inscriptions that have been brought to light in the experimental sciences.[94] Moreover, this result is in conformity with Jack Goody's analyses of the role that writing arrangements can play in redefining the relations among objects that are put into writing or transcribed.[95] Here we observe that the interactions are shaped by the deployment of de-monstrations. The latter continually redefine the status of the preceding proofs as well as the relations among them.

The examination of the entire set of messages shows that this work of redefinition leads to criss-crossed links among re-presentations of the de-monstrations. The production of new perspectives on the de-monstrations is the object of "chain reactions," or more precisely of chain translations. For example, the perspective proposed in the last message examined above is itself the object of other points of view, such as the following (again, the statements introducing new re-presentations are in bold type):

>t(A or not A) = t(true) by the last axiom
> ^^^
> => max(t(A), 1 - t(A)) = 1
> => t(A) = 1 or t(A) = 0.
>I think the problematic step is the first one, the
>use of the last axiom. I don't see the excluded

$$t(A \lor \lnot A) = t(A) + t(\lnot A)$$

[93] See especially Collins, *Changing Order.*

[94] See especially Latour, *Science in Action.*

[95] See Jack Goody, *The Domestication of the Savage Mind* (Cambridge: Cambridge University Press, 1977). See also Brian, *La mesure de l'État*; for the history of mathematics, Brian's text offers another illustration of the fruitfulness of Goody's methodological tools. See also Thomas Crump, *The Anthropology of Numbers* (Cambridge: Cambridge University Press, 1990).

>middle coming in here. Whether the use of the last
>axiom is justified depends indeed on the
>interpretation of logically equivalent.

   Um, I'm new at this, so maybe I'm putting my foot in it, but **it looks to me
like your problem is in the underlined section.** That formula is only correct if
$t(A) = 1$, which is far from a given. If, say, $t(A) = .99$, then $t(\text{not } A) = .01$, and
$t(A \text{ or not } A)$ will never equal $t(\text{true})$.

   All I see here is that if you put binary values into fuzzy logic, you get binary
values back out again.[96]

In this excerpt, the vision of an "error" in Elkan's proof is displaced
onto the level of a new inscription. Whereas certain participants earlier
saw in the underlined equality (and pointed out to their readers) a formulation of the principle of the excluded middle, this writer sees it (and
shows it to his readers) essentially as a "numeric constraint" that does
not permit the manipulation of fuzzy propositions.

   In looking at a teaching situation we saw how a rather large array of
manipulations and skills was associated with the exercise of logic; here
again we find recourse to several different types of manipulations, even if
the latter are sometimes perceived by the participants in the debate as
incommensurable. On this basis, we can grasp once again the importance
of taking into account the translations and distinctions produced by the
actors (an approach I have advocated in preference to the Bloorian approach to logic).[97] In fact, if I had considered that the author was bringing
the principle of the excluded middle to bear here, while invoking a structural homology between numeric and propositional manipulations for my
own part, we could not have fully seen the object of an opposition that
motivates the production of a new de-monstration.

   Through this new example, we observe that the production of messages
may be analyzed by following the series of contradictory or differential
visions that are exhibited by the actors. This methodological point is only
the corollary of an important result: the work of those evaluating Elkan's
proof consists in fact in producing other de-monstrations, which can then
be subjected to the same type of scrutiny as the original de-monstration.[98]
Thus one must not distinguish artificially between the activity of a critic
and that of the author of a de-monstration. The participants in the debate
adopt these two approaches simultaneously.

[96] D. Hefferman, message 834.
[97] Bloor, "Polyhedra" (1982).
[98] Let us recall that this situation is described by De Millo, Lipton, and Perlis as the
relatively infrequent case in which a proof is the object of scrutiny, use, and translation by
other researchers.

In a sense, the actors are at once constructionists and deconstruction-ists. The construction of a new de-monstration is achieved in fact through the deconstruction of previous de-monstrations and appearances. The critics' de-monstrations are subjected to the same fate as Elkan's proof: the appearances they put on display can be deconstructed in favor of oth-ers. As a result, the debate proceeds fundamentally from a constant dis-placement of scrutiny, according to a rhythm established by researchers as they bring to light either the existence or the absence of links between the various stages in the de-monstrative itineraries and the recourse to one logical principle or another; these instances of bringing to light are of course correlates of the deconstruction of the realities of the earlier appearances, for which the authors substitute new ones.

Moreover, a detailed study of the way the protagonists' viewpoints are sequentially linked shows still more clearly than before that the question of the correctness of Elkan's proof does not play a central role in the debates. The examination of an excerpt from one of the messages gives a good illustration of this phenomenon. We have already met its author, E. Ruspini. Not content with showing the absence of correctness of Elkan's proof, Ruspini attempts instead to convince his readers of the triviality of the result to which that proof leads. To this end, he proposes a very concise de-monstration of what he is presenting as the result of Elkan's work:[99]

> One does not need to resort to . . . Elkan's lengthy proof to see that if all tautolo-gies in the classical propositional calculus extended to fuzzy logic, then the only possible truth values are zero and one. To see this in a rather straightforward fashion, note that application of the axioms of fuzzy logic to the formula above leads to $\max(t(\text{alpha}), 1 - t(\text{alpha})) = 1$, from which it follows immediately that either $t(\text{alpha}) = 0$ or $t(\text{alpha}) = 1$.[100]

The object of this rapid de-monstration is clear. By showing that it is possible to obtain in two lines what he presents as constituting "in fact" Elkan's real result, Ruspini shows that the theorem is not the rare product of a difficult and particularly astute undertaking. By producing this new de-monstration, Ruspini modifies the significance of the original proof, among others, by displacing his readers' gaze from that proof to its au-thor. He makes Elkan look like someone who has mobilized dispropor-tionate means to demonstrate his theorem. Thus Elkan is made to appear clumsy rather than elegant.

---

[99] The use of a new de-monstrative schema in fact leads the author to propose a *transla-tion* of the result. Lakatos's study of the controversy over Euler's theorem offers other in-stances of redefinition of the results and mathematical objects to which the adoption of a new demonstrative procedure leads. See Lakatos, *Proofs and Refutations*.

[100] E. Ruspini, message 816.

However, Ruspini's approach is also based on his normative and per-formative proffering of priorities in the exercise of de-monstration. On the basis of his own skills, the author privileges the display of certain elements in relation to others. By indicating what must and what must not be shown, Ruspini is in fact negotiating the nature of the demonstrative practice in terms of what it must retain as important or disregard as superfluous.

We shall see later on that one of Elkan's responses to Ruspini's de-monstration and its analogues lies in the production of new texts that point to the existence of these de-monstrations in an elliptical way. Elkan will modify their primary signification by stating simply that other proofs of his theorem exist, thus confirming the correctness of his result. The de-monstrations of Elkan's critics, and that of Ruspini in particular, thus meet the same fate they dealt to the original proof. They are objects in turn of a series of de-monstrations that subjects them to various translations.

Indeed, certain writers seem to have anticipated that at least some of the participants would not read all of the de-monstrations addressed to the forum carefully, that they would adopt a viewpoint based on a rapid reading of a limited number of these de-monstrations. As a result, these same writers seem to be seeking opportunities to optimize their chances of being read in order to win in the debates; they are thus adopting a strategy of de-monstrative presence, seeking sometimes to maximize their visibility in the forum by multiplying their contributions, and trying to ensure that their own de-monstration is as visible as possible among the proliferating substitute proofs.

To support this hypothesis, we need to analyze the way the de-monstrative strategies of certain participants fully integrate an effort to make the proffered results visible. We must note in particular that the work of de-monstration derives its impact and its limits precisely from the framework of the participants' material and temporal access to the inscriptions produced, and that the question of this access is amply taken into account by at least some of the actors.

Making Certain De-monstrations Maximally Visible

Many messages stress the various material and above all temporal difficulties the actors encounter as they seek to gain access to Elkan's proof and to previous de-monstrations posted on the forum. These interventions amount to appeals for truncated citations of the texts in question and for the production of new de-monstrations. The accumulation of de-monstrations addressed to the forum thus consists in part of messages that attempt to respond explicitly to these appeals. To assess both the problems of access invoked and also the powerful invitation to the production of new de-monstrations included in these messages, it will be useful to examine a few excerpts:

> I've only *skimmed* the paper so far but it looks fairly damning of FL for productive engineering purposes. Is anyone aware of Elkan's work and if there are some opposing viewpoints that specifically address his concerns?[101]

> Elkan's argument is a straw horse, from what you have said of it. . . . I would like to see what he has to say. The argument about complexity is especially intriguing since Zadeh invented FL to deal with complex systems. So where can I get this document?[102]

> I don't know anything about Elkan, some of his complaints may or may not be true.[103]

> I myself have not had the time to look into Mr. Elkan's paper but from the discussion it appears he made exactly the mistake that I thought he had.[104]

> I haven't read Elkan's paper yet (it's only available in postscript and I have no means of reading a postscript document; a friend is going to print me a copy RSN), but if this is what he's saying then he's missed the point somewhere.[105]

> Hi, I haven't been following reactions to Elkan's paper too closely. But I did go through his paper.[106]

> There was no other meaningful discussion of the rest of his paper which presents his opinion that future attempts to apply fuzzy logic as an approach to the area of heuristic control would be limited by the same problems that other researchers have with non-fuzzy techniques (I don't have the paper in front of me so I can't summarize his presentation at this time.)[107]

> Unfortunately I missed this discussion and can't recover the "back numbers" of this newsgroup. If anyone still has their comments on file, I would be grateful if you could email them to me. Thanks.[108]

> Hello,
>   I have a question: Is Elkan's paper on the reducibility of Fuzzy Logic available on the net? Thank you for your help.[109]

---

[101] D. Lawler, message 764.
[102] J. Wiegand, message 767.
[103] W. Dwinnell, message 769.
[104] J. Wiegand, message 832.
[105] D. Hefferman, message 834.
[106] S. Rangaswamy, message 1051.
[107] D. Lawler, message 853.
[108] P. Jackson, message 889.
[109] Allen, message 1778,

Hi Everyone,

I am currently doing a project on fuzzy logic, and would like to know whether anyone here has kept some of the comments for/against the paper Professor Elkan wrote on the reducibility of Fuzzy Logic? If so can you post them or email them to me? If not can some kind soul please give some opinions on it? By the way, are there any other papers that attack the credibility of fuzzy logic? Thanks in advance.[110]

Does anyone have a summary of the responses to Elkan's paper "The Paradoxical Success of Fuzzy Logic," particularly with respect to the proof that fuzzy logic reduces to bivalued logic? Also, have there been any papers or articles published that address the issues raised in Elkan's paper?[111]

I include in this and the following mail some of the posts regarding this paper (one is by Charles Elkan himself, and he states in it where to get his paper via anonymous ftp).[112]

These messages convey a series of avowals of partial readings or non-readings of Elkan's original proof and its posted substitutes. They contain a large number of requests for information about how to gain access to the various texts. They also ask that the proofs be sent to the forum again, and that new commentaries or substitute proofs be produced.

We must recall that the messages addressed to comp.ai.fuzzy remained on display in this forum only for a week. The messages devoted to Elkan's theorem, produced over a much longer period of time, were thus not accessible simultaneously. The actors who had not mastered the skills required to consult the forum's archives, and especially those who lacked the necessary time or motivation to find messages buried in the quite voluminous archives, thus had no better option than to appeal to the "regular" participants in comp.ai.fuzzy who had kept their own records of the exchanges over time.

A detailed examination of the preceding series of excerpts reveals the multiplicity of material constraints that weigh on access to the demonstrations. Certain authors assert that they have not had the time required to follow the debates in the forum (in detail), or to read Elkan's article (carefully). Others state that they do not have the article *in front of them*, so they cannot comment on it in real time. Still others declare that they cannot read a certain text by Elkan, accessible on the Internet through an FTP command, inasmuch as it is expressed in a format that is unreadable for those who do not have the resources or the skills needed to print out a hard copy.

---

[110] V. Chaojirapant, message 1805.

[111] M. G. Cooper, message 1870.

[112] W. Slany, August 31, 1993, cited in message 1823.

It is clear that the accumulation of these problems of access to texts left the path wide open for those proposing to substitute new de-monstrations for the proofs published earlier, and in particular for Elkan's original proof. At the sight of the difficulties displayed and the requests expressed, the actors who had privileged access to Elkan's article and who had taken the time to follow the debates in the forum were in a position to play a key role in the collective dynamics of formation of points of *view* on the original proof, by proposing new de-monstrations or by reformulating the contents of the texts previously addressed to comp.ai.fuzzy. The messages that were forwarded or copied most often in the forum—that is, the ones whose visibility was greatest in practice—were also those that had the greatest chance, all else being equal, of imposing their viewpoint in the long run. In other words, the impact and the limits of the de-monstrations were in large part a function of the material and temporal conditions of the readers' *access* to the various texts.

In fact, the analysis of the entire set of messages shows that, under the conditions of this material economy of access to the texts, certain substitutions for Elkan's proof had at least as much impact on the representations that the participants in the forum constructed of the proof as the original text itself. To be sure, it is not easy to define or pinpoint the protagonists' adoption of particular viewpoints on Elkan's proof. But it is not unreasonable to attempt to define and grasp these viewpoints simultaneously in light of the participants' expressions of opinions, sometimes conditional, based on messages written by third parties.

In the excerpts cited above, a number of writers thus declared that their appreciation of Elkan's proof was based solely on secondary sources (primarily messages addressed to the forum). Only a few voiced some doubt as to the faithfulness of the translations proposed. The role played by the proffering of substitutes in the formation of viewpoints on Elkan's proof can also be grasped from messages expressing a modified or solidified opinion on that proof following the reading of a text written by a third party. The following example is a good illustration of this phenomenon:

$$t(\text{not}(A \text{ and } \text{not}(B))) = t(B \text{ or } (\text{not}(A) \text{ and } \text{not}(B)))$$

This statement was worrying me too. I was wondering if it really should hold in fuzzy logic. There are other statements in classical logic that don't work in fuzzy logic. It would seem the answer is no, it wouldn't. I wasn't sure if I was missing something though, but I think Loren's post illustrates the mistake in it.[113]

The author of this text clearly affirms that reading a message on comp.ai.fuzzy *reassured* him about his idea that Elkan had made a mis-

---

[113] S. Kroger, August 2, 1993, cited in message 1823.

take in a specific passage of his proof. The solidification of the viewpoint at issue can be grasped in the description of the shift from a phase in which the author *wasn't sure* about a statement or about his own ability to have detected an error in it, to a phase in which he now *thinks* (although without absolute certainty) that the message he cites has *brought* this error *to light*.

In this connection, it is important to stress the extent to which the space of interaction constituted by an electronic forum of this sort is a privileged terrain for the sociologist. It allows researchers to retain (and later to display) the traces of the processes through which representations about a proof are formed, traces that would be difficult to obtain from other sources. In particular, being able to apprehend phenomena as subtle as the dynamics through which viewpoints are solidified is a crucial aspect of the study of the debates that animated this particular newsgroup.

Before concluding this description, I must again point out the essential role played by recourse to a different type of device in the formation of representations concerning Elkan's proof. We have seen that certain authors had considerable margins for maneuver, allowing them to ensure higher visibility for their own messages. In fact, their competing individual undertakings were sometimes accompanied by actions carried out by groups. The predominance of certain messages (in terms of visibility on the Internet) was ensured on the one hand by repeated references to the latter (their interest was emphasized on many occasions in an organized way). It was ensured on the other hand by the creation of a device that allowed ready and uninterrupted access to these same messages, making them by that very token more visible than other texts.

This arrangement consisted in placing "selected" texts in a database linked to the network, a database whose electronic address was communicated on comp.ai.fuzzy's "frequently asked questions" (FAQ) page. This amounts to saying that the forum's managers had in their hands an important instrument for structuring the viewpoints on Elkan's theorem that were being forged. The following two excerpts illustrate this point clearly:

> Please read especially the enlightening and very well written evaluation by Enrique Ruspini (sent in the next mail). . . . Again, please read especially the enlightening and very well written evaluation by Enrique Ruspini (first message in this mail).[114]

> Two responses to Elkan's paper, one by Enrique Ruspini and the other by Didier Dubois and Henri Prade, may be found as ftp.cs.cmu.edu:/user/ai/areas /fuzzy/ doc/elkan/response.txt.[115]

[114] W. Slany, August 31, 1993, cited in message 1823.
[115] FAQ, message 2072.

In the first message, from which I have cited only a short excerpt, several elements help give Ruspini's text visibility: the reproduction of its contents, of course, but also the insistent reference to this content in a tone of advocacy at the beginning and end of the message. In the second excerpt, which is from one of the forum's FAQs, we note that a supplementary element helps increase the prominence of Ruspini's text, along with that of a message written by two other authors, Didier Dubois and Henri Prade:[116] the communication of the electronic address of a database where these two texts were stored.

From these examples, as from the earlier ones, we can see to what extent the struggle to persuade could be waged by way of a struggle to make texts visible. In the end, the formation of viewpoints toward Elkan's de-monstration appears amply structured by the *advancing* of some de-monstrations that are presented as substitutes for the original proof, and whose authors were able, most importantly, to have access to Elkan's text.

In other words, we observe—and this point is essential—that the constitution of the representations of Elkan's de-monstration does not depend on aggregating individual examinations of the text of the proof. It stems rather from the appreciation by readers who are more or less distant, in the material and especially the temporal sense of the term, of a magma of parallel de-monstrations that are often antagonistic and highly unequal in terms of visibility. This in particular is why the nature of the recognition a theorem such as Elkan's was able to acquire after its publication could not be determined at the outset on the sole basis of an examination of the text of its proof: many other elements, whose diversity we are only beginning to glimpse, intervened in the collective dynamics of the formation of judgments on this proof.

At this stage, we have already arrived at a nonclassical representation of the discussions that a logical proof may stimulate. However, if the analysis ended here, readers might think that the controversy studied had essentially to do with the question of the correctness and value of Elkan's proof. They would doubtless conclude that they had been confronted quite simply with a debate free of any other considerations. Even if they were familiar with the abundant literature produced in the sociology of the experimental sciences,[117] exhibiting multidimensional exchanges, they might well think that the exceptional rarefaction of the stakes was characteristic of the field in which the debates were inscribed (that of logic).

---

[116] An excerpt from a contribution by these two authors was analyzed above, p. 93.

[117] The list of introductory articles and books on this topic is a long one. See especially Dominique Vinck, *Sociologie des sciences* (Paris: Armand Colin, 1995); Sheila Jasanoff, Gerald E. Markle, James C. Petersen, and Trevor Pinch, eds., *Handbook of Science and Technology Studies* (London: Sage, 1995); Mario Biagioli, ed., *The Science Studies Reader* (London: Routledge, 1999).

However, this is absolutely not the situation in which we find ourselves. We now need to examine various de-monstrative modes and related controversies associated with the "discussion" of this proof. It will be a matter of studying how other modes of de-monstration are deployed and put in the balance in the face of the formal de-monstrations, and how the actors are led to introduce hybrid registers in order to participate in the debates. Now that I am about to show its limitations, I should like to note that the artificial focus on formal de-monstrations that I have maintained to this point has at least one virtue: it clearly emphasizes the fact that not every debate over a logical theorem is necessarily a *purified* debate, that is, a debate involving purely formal considerations.

Chapter 4

EVALUATING THE CORRECTNESS OF A THEOREM AND THE PROPERTIES OF A LOGIC AT THE INTERSECTION BETWEEN SEVERAL DE-MONSTRATIVE MODES

BRINGING TO THE FORE THE PROPERTIES OF A LOGICAL SYSTEM IN TECHNOLOGICAL DEVICES IN ORDER TO CAST DOUBT ON THE CORRECTNESS OF A PROOF

IN THE INTRODUCTION to this work, I evoked the fact that Elkan put forward the "inconsistencies" of fuzzy logic in his article in order to denounce the paradoxical dimension of the success of its putative applications. This thesis gave rise to debates on comp.ai.fuzzy over the effectiveness, the nature, and the causes of such "success," all of which are indissociable from the exchanges bearing on the correctness and the value of the theorem. These debates were structured by recourse to heterogeneous de-monstrative modes. The authors mixed various considerations and registers to debate the overall set of properties—whether or not these were deemed "formal"—of fuzzy logic. Various ontological forms associated with that object were invoked (in particular, technical devices known as fuzzy) in order to guarantee the reality or the lack of reality of its properties.

The exhibition of properties of fuzzy logic in computer systems thus came to be opposed to the results of a symbolic demonstration, or else those properties were evoked, on the contrary, in order to underline the correctness of that demonstration. In other words, the same message could include intersecting considerations on Elkan's proof, its substitutes, and the causes of the working of a given electronic device. The way a particular machine worked could for example be attributed to quite general properties of fuzzy logic, and presented by that very token as calling back into question the correctness of Elkan's overall de-monstration (including the proof of his theorem regarding the "inconsistencies" of fuzzy logic).

The debates dealt in fact not only with the reality and the nature of the properties of fuzzy logic (*embodied*, for some, in technological devices), but also with the modalities of the staging of those properties in the texts of the various participants. One could then witness a sequential linking of constructionist and deconstructionist undertakings, the *appearances* of properties attributed to fuzzy logic in certain messages being deconstructed to the profit of others during later exchanges.

It is time now to study closely, with an eye to pinpointing their specificity, the de-monstrative registers weighing against each other, which were produced by these weavers of proof.[1] Let us begin by analyzing a first excerpt from a message that puts the properties of fuzzy logic in electronic and computer systems on display in the face of the critiques developed by Elkan:

>Most of the applications in the control area using FL
>have been too simple to date to show the inherent
>weaknesses of the technique.
 Like the fuzzy predictive controller that handles the Sendai trains? Or the Otis elevator scheduler that uses both predictive and interactive inputs? How complicated does it need to be? 100,000,000 rules?
>Once these weaknesses become apparent then FL will
>turn out to be just as problematic to use as other
>knowledge-based techniques (and he seems to imply
>possibly more so).
 Well, it IS problematic if you don't understand it. It has taken me well over three years to grasp the philosophical basis of FL. What other AI technique has the properties of universality and proximality? Just neural nets, and they are closely related to FL. The Japanese tried AI in their Fifth Generation project, and it got them nowhere. But now they lead the world in fuzzy patents. Comparison? Just the opposite of what Elkan is saying. . . . The properties of FL should be obvious to anyone who has investigated it with an open mind.[2]

This excerpt, written by J. Wiegand, is an example of calling Elkan's de-monstration back into question without going through a critique of the details of the symbolic proof of the theorem. We are dealing here with a form of reductio ad absurdum, which can be reformulated as follows: if fuzzy logic were the victim of formal inconsistencies, as Elkan's proof announces, no complex mechanism conceived with the help of fuzzy logic

---

[1] It is not a question here of revisiting the general forms of justification, as they are derived from the analysis performed by Luc Boltanski and Laurent Thévenot in *On Justification*, trans. Catherine Porter (Princeton: Princeton University Press, 2006). My approach is guided here by a distinct problematics: the attempt to pinpoint the specificity of the de-monstrative modes used in the framework of a debate over a logical theorem.

[2] J. Wiegand, message 767.

could work;[3] now, there exist complex mechanisms that work thanks to fuzzy logic; thus Elkan's de-monstration does not "hold up."

With regard to the way sophisticated fuzzy systems work, Wiegand insists in fact more precisely on the absence of *plausibility* of the formal inconsistencies of fuzzy logic advanced by Elkan. The exhibition of controllers whose working is attributed to fuzzy logic thus constitutes a first de-monstrative register short-circuiting the debates over the details of the proofs. Confronted with the de-monstration of the theorem, he stresses, finally, its *implausibility*.[4]

The de-monstrative mode in play stems clearly from a practice of *bringing to the fore (faire valoir)*. It consists, in the face of an enterprise of devalorizing fuzzy logic, of seeking to *bring* properties of fuzzy logic *to the fore* in electronic instruments. To this end, Wiegand exhibits commercial products emblematic of brands and of a technology, fuzzy logic in this case. The author presents these devices as literally *embodying* certain properties of fuzzy logic. The successful operation of controllers used by the Sendai train and an Otis elevator is thus *put into the balance* against Elkan's de-monstration.

Other de-monstrative registers are also used. Paralleling the reference to the operating principles behind trains and elevators, elements as heterogeneous as Japanese leadership in the realm of fuzzy patents and the overt "failure" of a research program in artificial intelligence mobilizing a logic with two truth values (the "fifth generation project") are evoked in order to exhibit very general properties of fuzzy logic, characterized as "philosophical" (like the property of "universality").

The author does not stop, however, on such a good path. He goes on to mobilize an additional de-monstrative register, one we have already identified in the course of the previous analyses: a *discourse about the self-evidence* of the properties of fuzzy logic taken as an object. This involves a de-monstrative procedure abundantly illustrated by the protagonists in the debates, and much earlier by authors such as Husserl.[5]

Wiegand insists on the ideal self-evidence of certain properties of fuzzy logic; then, to account for the fact that this self-evidence may not appear as such in Elkan's eyes, he invokes "contingencies" and, above all, "psychological resistances." For the author, such blindness may stem from a biased attitude that originates in the mind of its victim: a "lack of open-mindedness" would be its cause. Moreover, Wiegand evokes the long pro-

---

[3] Let us note that the author does not challenge the claim according to which formal inconsistencies of fuzzy logic could fail to manifest themselves in simple technological devices (ones that work owing to that logic).

[4] For other uses of the "plausible" in mathematical de-monstrations, see Brian, *La mesure de l'État*, p. 60.

[5] See once again Husserl, *Logical Investigations*, pp. 46–47 and 63–64.

cess of decantation that may prove necessary to accede to that self-evidence, however immediate it may be from the ideal viewpoint. That affirmation is shored up by *testimony*, since the author makes much of the three years he needed to apprehend it as such.

Thus we find a de-monstrative mode mobilized in other messages being deployed once again at the end of a de-monstration. By this very token we can appreciate the widely distributed character of its use. Let us now study Elkan's response to this message (and others) on the electronic forum, analyzing with special care the set of de-monstrative registers he mobilizes for the occasion:

> FUZZY LOGIC IS USEFUL IN ENGINEERING The paper is not meant to be "damning of FL for productive engineering purposes." Fuzzy logic has been and will be very successful in heuristic control applications, for example for Sendai subway train braking and for elevator speed control. However these systems are small compared to other knowledge-based systems. They use less than 100 rules, compared to many thousand for many expert systems. The fact that Japan leads the world in fuzzy patents and in fuzzy controllers has several plausible non-mystical explanations: (a) Japan leads the world in manufacturing high-technology consumer products. These are the largest natural application area for fuzzy controllers.(b) Japan leads the world in patents overall. The number of patents per year that a company chooses to take out is a business decision influenced by many concerns and is not perfectly correlated with the company's overall success in research OR development. (c) Fuzzy logic controllers are engineered in an iterative, heuristic process of incremental improvement. This is congruent with traditional Japanese strengths in incremental quality improvement. (d) The reasons fuzzy controllers work well are that they (1) are rule-based and (2) have many tunable numerical coefficients. More applications outside Japan may use these two features without using the keyword "fuzzy."[6]

Here, Elkan accompanies his original de-monstration by translating it, that is, by producing new de-monstrative registers. First of all, he defends an affirmation that constitutes one of the keystones of his overall de-monstration: the number of rules used by fuzzy systems is too limited for their proper functioning to be disturbed by the "formal inconsistencies" proper to fuzzy logic. To this end, the author evokes once again the operation of devices such as the braking system of the Sendai train and the automatic speed controllers of certain elevators. The author attributes to these controllers the use of a number of rules that is also quite limited.

Beyond this, Elkan refuses to attribute the "success" of fuzzy systems to properties of fuzzy logic. He deconstructs that causal chain in order to make a new one. He *associates* the large volume of fuzzy controllers

[6] C. Elkan, message 794.

produced in Japan with a general economic dynamics, Japanese *leadership* in the realm of production of high-technology goods (the cause of the success evoked is thus displaced a first time: it stems from a more general phenomenon). Then he *dissociates* the data concerning the volume of patents held by Japanese companies from the success of these companies in the realm of research and development. He then *associates* the creation of fuzzy controllers in Japan to a Japanese cultural specificity in working methods: the latter would rest on step-by-step improvements in the devices that are being developed. Finally, he *associates* the successful operation of fuzzy controllers not with specific properties of fuzzy logic, now, but with the fact that they are developed on the basis of rules and numerical coefficients capable of being optimized.

Elkan thus deploys a set of heterogeneous de-monstrative registers to defend his original de-monstration: socio-techno-economic considerations on the market in high-technology goods, putative expertise in the organization of labor, production, and management in Japanese businesses, as well as a "paternity trial" bearing on the causes of the proper functioning of fuzzy controllers. Such a trial constitutes the counterpart of a practice of bringing to the fore. It consists here in revealing how fuzzy devices actually work, and showing that these mechanisms do not depend on properties *specific* to fuzzy logic. Elkan ends up advocating that the *emblem* "fuzzy logic" no longer be used to account for this.

In the face of this work of respecification of the causal chains, several participants in comp.ai.fuzzy plunged into analogous paternity trials. Some sought to bring the emblem of fuzzy logic to the fore once again by presenting the properties of fuzzy logic as embodied in technological devices. The electronic forum was thus the theater for a series of appearances, not solely those of logical principles like the ones we saw earlier, but also those linked to the construction and the deconstruction of the properties of fuzzy logic. The two excerpts below offer a good illustration of this phenomenon.

> >(d) The reasons fuzzy controllers work well are that
> >they (1) are rule-based and (2) have many tunable
> >numerical coefficients.
> IMHO, the reason fuzzy controllers work well is because the rules are FUZZY.
> >More applications outside Japan may use these two
> >features without using the keyword "fuzzy."
> Darn tootin. First person to think of a better surrogate word than "intelligent" gets a cookie from the marketing department.[7]

[7] S. Moses, message 795.

>the reason fuzzy controllers work well is because the
>rules are FUZZY.

Hear, hear! And because they possess fuzzy attributes such as rule overlap and aliasing. I've written a fuzzy controller where the placement and weights of membership function evaluator triangles is easily user-adjustable . . . Get rid of rule overlap and believe me, one can tune the numerical coefficients to one's heart's content and never get the system operating with anything approximating its original smoothness and efficiency.[8]

In the first of these messages, the author makes fuzzy logic *reappear* as the first cause of the proper functioning of controllers called "fuzzy." The graphic technology mobilized on this occasion consists in juxtaposing a quote from Elkan's text and the emblem "fuzzy logic." With the help of capital letters and an ironic tone, this latter is thus *displayed* at the heart of the new causal chain.

In the second message, the author goes a step further in putting forward the determining character of the specificities of fuzzy logic to account for the successful operation of "fuzzy" devices. To this end he adopts the de-monstrative register of *testimony*, relating his personal experience in developing controllers. By a visual procedure of graphic juxtaposition, the emblem "fuzzy" is associated with various practices presented as specific to implementing the logic in question—in particular, regulating and adjusting the parameters of a controller by the combined application of weighted rules that may be represented graphically by overlapping triangle functions.[9]

We see how, in the course of the exchanges, the properties of fuzzy logic can just as well disappear as reappear in material devices represented textually, as the various de-monstrative registers are deployed. The study of the foregoing excerpts shows how the transformations of representations of the way so-called fuzzy controllers work, in essence, are linked. This itinerary is organized by inscriptions that punctuate the displacement of gazes on the reality, or lack of reality, of the properties of fuzzy logic. Still, in order fully to grasp the way reversals of viewpoint on Elkan's de-monstration are elaborated in confrontations between various de-monstrative registers, it will be useful to analyze one additional excerpt. In the passage that follows, the author, E. Ruspini, makes certain properties of fuzzy logic contested by Elkan reappear:

---

[8] R. Teasdale, message 797.

[9] The analysis of the way controllers are developed would take us too far from our topic. On this topic, see Bernadette Bouchon-Meunier, *La logique floue* (Paris: Presses Universitaires françaises, 1993), pp. 100–114; R. R. Yager, ed., *Fuzzy Sets and Applications* (New York: John Wiley and Sons, 1987).

Perplexed by the fact that fuzzy logic has attained considerable success in applications despite this lack of formal soundness, Elkan devotes most of his paper to account for this paradox. His conclusion is that, so far, the inadequacies of FL "have not been harmful in practice because fuzzy controllers are far simpler than other knowledge-based systems." This is an extraordinary assertion because fuzzy logic has been successfully applied to synthesize controllers for rather complex systems (e.g., Sugeno's application to helicopter control). Furthermore, some of these controllers have considerably more elaborate architectures than those mentioned by Elkan in his paper. At any rate, even if all the fuzzy controllers developed to date were simple, as Elkan thinks, it is arguable that reliance on a single inferential step would make them less prone to failure. . . . Elkan is also misled by the relatively small size of rule-sets in fuzzy controllers. It is generally agreed that the compactness of fuzzy knowledge-bases is the consequence of their inherent ability to describe complex systems by introduction of powerful approximation tools into an inferential framework. By contrast, conventional inferential techniques generally require, in a control problem, the specification of a control value for each conceivable value of the state. Elkan confuses what is a desirable property of fuzzy systems with a supposed lack of important applications of the technology.[10]

In this excerpt, the writer attempts in turn to cast doubt on the plausibility of Elkan's results by contesting the compatibility of two of Elkan's assertions: on the one hand, fuzzy logic is a victim of "formal inconsistencies," and on the other, "fuzzy" devices work well. Ruspini challenges Elkan's explanations for the apparent "paradox," which consists in the first place in putting forward the simplicity of the "fuzzy" controllers developed up to this point. In contrast, Ruspini stresses their complexity, in particular by exhibiting a new object, a helicopter whose operation he attributes to fuzzy logic.

But after weighing this new object against Elkan's de-monstration in order to highlight the properties of fuzzy logic, he attempts to orchestrate a new displacement of the reader's gaze by adopting a *techno*-logical register in his own text. The small number of rules mobilized to develop fuzzy controllers, after having been associated by Elkan with an ideal *limitation* of fuzzy logic, are brought in here to illustrate an *intrinsic property* of that logic, of its "descriptive power." Ruspini goes on to explain that these same devices are easier to build.[11]

In the context of this new de-monstration, a specific fact *illustrates* a new reality. The small scale of the architecture of fuzzy controllers is no longer the source of the absence of manifestations of *problems specific* to

---

[10] E. Ruspini, message 816.

[11] I shall not analyze in detail the practices to which the author refers to explain how controllers are built, in particular the ways their states may be specified. Such an undertak-

fuzzy logic. It becomes the consequence of a *property* of this logic, that of making it possible to manage complex systems.

Elkan's original de-monstration is thus submitted to a series of rebounding paternity trials bearing on the mechanisms underlying the operation of fuzzy devices. Here, "small" becomes "compact"; a helicopter is mobilized literally to fly to the rescue of the consistency of fuzzy logic. The unfolding of the de-monstrative registers that underline the embodiment of properties of fuzzy logic in material devices appears in much the same way that symbolic proofs were being advanced. We rediscover the ballet of constructionist and deconstructionist approaches linked to a work of association and disassociation, appeals to self-evidence, sequences of appearances and disappearances, work on exemplarity/exemplification that is synonymous here with the notion of bringing to the fore. The disagreements over what certain protagonists seek to show and what others see are just as important, and the points of agreement just as limited.

With a case like this, we see how negotiating the "logical" character of the operation of technological devices[12] could not be fully apprehended without a systematic analysis of the de-monstrative registers mobilized on this occasion. In fact, several factors led me to limit my analysis to the processes of *attribution* (or nonattribution) of properties of fuzzy logic to the devices being highlighted: given the considerable volume of black boxes deployed, the fact that a good number of these remain closed at the end of the foregoing descriptions, as well as the highly solidified disagreements expressed in the framework of the paternity trials. In the absence of other elements, it was hardly possible to adopt a different approach, and in particular it was out of the question to reach a conclusion as to the ontological characteristics of the controllers.[13] We must nevertheless note that this methodological option has played a major role in enabling us to grasp the way heterogeneous de-monstrative resources, corresponding to various objects, practices, and skills, could be used to confront

---

ing would take us too far from the project at hand. On this topic, see H.-J. Zimmermann, *Fuzzy Set Theory and Its Applications* (Dordrecht: Kluwer, 1985).

[12] For an analysis of a similar dynamic, see MacKenzie, "Negotiating Arithmetic, Constructing Proof."

[13] Moreover, the descriptions of these devices generally mobilize such a large number of black boxes that the actors frequently limit themselves to the discourses of attribution of properties proffered by the "inventors," though these same actors may do so in the conditional mode. One illustration of this phenomenon can be found in a report by an "expert" working for an institution called the Office of Naval Research–Asia; the report deals with the operation of Sugeno's helicopter (message 1779). Despite all his efforts to document his report, his interviews with Sugeno, and his on-site observations of the helicopter's operation, the author does not manage to alleviate all the uncertainties that weigh on the causes and modalities of this operation, and finds himself obliged to cite the "inventor" without being able to defend or deny the value of his assertions.

proofs expressed with the help of ordinary logical symbols, so as to underline or contest their correctness.

"Fuzzy" machines ultimately appear as operators for advancing the properties of fuzzy logic, on the same basis as symbolic proofs.[14] We see here that the formalist exercise, in the sense in which we defined it earlier, does not produce unilateral and large-scale effects on a wide range of activities without being affected by those activities. Our observations do not call for such an architectonic representation of knowledge, quite the contrary. The status granted to a logical theorem depends here in part on representations of the operation of technological devices produced in places that are not themselves centers of calculation, but research centers that develop controllers and factories that produce them; it also depends on how the actors represent the volume of fuzzy patents registered in Japan, the market in high-technology goods, and the organization of labor, production, and management in Japanese firms.

Whereas we might have thought that the production and the discussion of a logical theorem proceeded in all cases from a quite well delimited genre, the controversy is unfolding here in multiple directions. To be sure, that should not surprise us, given the identity of the participants. The researchers in artificial intelligence who are taking part in the debates constitute *mediators* capable at once of describing the way technological devices are built, using skills linked to the manipulation of ordinary logical symbols, and carrying out translations between the corresponding registers.[15] The splintering of the de-monstrative registers is in addition closely linked with the involvement of fuzzy logic in multiple activities, in the same way that the exercise of analysis in the eighteenth century, as Éric Brian shows, stemmed from operations of decomposition that structured practices belonging to diversified activities.[16]

[14] This result must not be considered in any way exceptional. The "rise" of fuzzy logic in the first half of the 1990s was implemented in a crucial fashion by the exhibition of devices designated "fuzzy." This type of undertaking went a long way toward ensuring the spread of research in the field. On this topic, see Claude Rosental, "Histoire de la logique floue: Une approche sociologique des pratiques de démonstration," *Revue de synthèse* 4, no. 4 (1998): 573–602. In comparison, see, in Warwick, "Laboratory of Theory," how calculation tables and machines constituted vectors for the development of mathematical analysis in nineteenth-century England.

[15] For some elements of the history and sociology of research in artificial intelligence, see Jean-Pierre Dupuy, *Mechanization of the Mind: On the Origins of Cognitive Science*, trans. M. B. DeBevoise (Princeton: Princeton University Press, 2000); Pierre Lévy, *De la programmation considérée comme un des beaux-arts* (Paris: La Découverte, 1992).

[16] Such was the case in particular for the procedures of integration, demonstrative *compositions*, methods of administrative counting, and the elaboration of tables of types of knowledge. Here again, see Brian, *La mesure de l'État*.

In the end, we again encounter the "impure" objects (involving science, technology, and society simultaneously) of sociological studies dealing with the experimental sciences and their controversies.[17] However, it is appropriate to stress at this point that we also find participants in these debates on the forum adopting an approach often brought to light in the analyses of scientific controversies: (re)personalization, by the actors, of the positions defended by the various parties. This phenomenon runs contrary to a highly sedimented image based on the rejection of ad hominem arguments:[18] according to this image, in a debate in logic, only the arguments count and not the individuals who advance them.

Personalizing the Debates in Order to Evaluate the Correctness of a Theorem

It is particularly striking to observe to what extent the question of the institutional identification of certain actors can preoccupy the protagonists in the debates, if one compares this situation to the emblematic figure of the logician as it is stabilized in the rejection of the ad hominem argument. The question of Charles Elkan's institutional positioning and that of a specialist in fuzzy logic, Bart Kosko, is in fact very present in the course of the interventions, as the following selection of excerpts attests:

It's interesting to note that he [Elkan] comes from the same school as Bart Kosko. I wonder if the two of them have compared notes or are having an ongoing dialogue?[19]

While I am not familiar with Dr. Elkan's work, I am pretty sure that Bart Kosko is from USC. Did you mean that Elkan is from USC also? If not, do you know what department Elkan is in at UCSD?[20]

I think Kosko is out of UCB. From what I have heard about Kosko, it would be more like a monologue.[21]

[17] See especially Vinck, *Sociologie des sciences*.
[18] Literally, "against the man." An ad hominem argument consists of replying to an argument by attacking or appealing to a characteristic or belief of the person making the argument, rather than by addressing the substance of the argument. See Douglas N. Walton, *Arguer's Position: A Pragmatic Study of Ad Hominem Attack, Criticism, Refutation, and Fallacy* (Westport, Conn.: Greenwood Press, 1985).
[19] D. Lawler, message 764.
[20] D. Wahl, message 765.
[21] J. Wiegand, message 767.

Bart Kosko is at USC, and does anyone have his e-mail address?[22]

Bart Kosko is at the University of Southern California (USC) in Los Angeles. I am at the University of California, San Diego (UCSD). The only connection is that he got his M.S. at UCSD. Charles Elkan.[23]

I had previously mislocated B. Kosko at UCSD. Dr. Elkan was kind enough to point out that Dr. Kosko is currently with USC.[24]

These excerpts show that in more than one respect we are not dealing with a simple debate about ideas. Indeed, not only does the unfolding of the controversy involve a multiplicity of material mediations, as we have already seen, but it is also based on the direct implication of the protagonists' identities and institutional affiliations.

While we have been able to identify up to this point a multiplicity of de-monstrative registers associated with heterogeneous modes of exhibition, this display of "persons" can be viewed as a de-monstrative register in its own right. This register in fact enters into the dialectical process of strengthening or negating the reality of the properties of fuzzy logic, as it can be added to the scales with other de-monstrative registers. It offers the protagonists in the debates a supplementary means to evaluate the solidity of Elkan's de-monstration: in the case in point, testing the depth of its author's dialogue with one of the leading figures in fuzzy logic.

Bringing to light such a practice in the debates thus both underlines and extends other sociologists' descriptions of phenomena associated with the repersonalization of statements in situations of controversy in the experimental sciences. Bruno Latour in particular has succeeded in showing that the strengthening of experimental "facts" could pass through a long decanting process during which the experimenter's claims would become increasingly impersonal, and that the opening of a controversy, by re-opening black boxes, could lead to repersonalizing the claims.[25] Simon Schaffer, for his part, has shown that some public experiments were conceived so that the demonstrator might be as transparent as possible, and that the claims made could be challenged by bringing the experimenter back before the public eye.[26]

My own research into the modes of teaching logic led to comparable results. Divergences bearing on logical statements sometimes led to open conflicts between teachers and students. The radicalization of the exchanges was then accompanied on both sides by a personalization of the

[22] J. Craiger, message 770.
[23] C. Elkan, message 794.
[24] D. Lawler, message 853.
[25] See Latour, *Science in Action*.
[26] Schaffer, "Self-Evidence."

adverse positions and by denunciations of pathological attitudes. When such a stage was reached, it represented a point of no return in relation to the dialogue and marked, de facto, the end of the debates.[27]

To be sure, many sociologists who do not take science as their exclusive object have looked into the dynamics of impersonalization and personalization of statements and behaviors: these would include Durkheim, for whom "impersonal reason is only another name given to collective thought,"[28] and also Bourdieu, when he develops his notion of objectivization.[29] But our point of entry into the analysis of these phenomena arose here from a specific problematic: it was first and foremost a question of studying how the personalization of debates could constitute a particular de-monstrative register capable of being weighed against other registers in the context of a controversy.

To pursue this analysis, we need to examine the extent to which the exhibition of actors in the context of exchanges on comp.ai.fuzzy did not operate simply through the identification of *individuals*. Our study of the messages shows in fact that several participants challenged the demonstrations of their adversaries by pointing to conflicts between *groups* of actors. Thus we are often dealing with an authentic de-monstrative resource, with a register destined to short-circuit the debates over the details of the proofs, as we shall now see.

TRYING TO NEUTRALIZE A PROOF BY INVOKING GENERAL ANTAGONISMS

Several messages addressed to the forum associate the writing of Elkan's de-monstration with the resonance of a generalized antagonism to fuzzy logic among researchers in artificial intelligence. The nature and the declared objectives of the associations made in this way are quite varied.

First of all, certain messages present Elkan's de-monstration as constituting *essentially* the mark of a more global intent to denigrate fuzzy logic in artificial intelligence. Such messages do not position themselves in relation to the question of the correctness of this demonstration. By this very token, they identify Elkan's article with, or even sometimes implicitly reduce it to, that antagonism.

As an illustration, let us examine an excerpt from a message that short-circuits the debates over the details of the proofs by adopting a specific

---

[27] Rosental, "De la logique au pathologique."

[28] Durkheim, *Elementary Forms of Religious Life*, p. 446.

[29] See Pierre Bourdieu, *Homo Academicus* (Stanford, Calif.: Stanford University Press, 1988 [1984]), pp. 15–19.

de-monstrative register. This register consists in contributing *testimony* bearing on the existence of a distancing attitude on the part of researchers in artificial intelligence toward research carried out on fuzzy logic. The author pronounces himself ready to associate this attitude with a situation of cold war, one that Elkan's article itself would simply *reveal*:

> It's my understanding that Dr. Elkan's paper received an award at the last AAAI conference where it was presented. Several people I've discussed this with have expressed surprise at this award (although I'm sure that Dr. Elkan is fully capable of making a fine presentation and no one has observed otherwise). Comments were made about a prejudice against fuzzy logic in general being part of current AI thinking. I've been to a number of AI conferences where fuzzy concepts were discussed (many with Zadeh present), but it does seem to me that fuzzy logic has not been well accepted in the "main stream" AI community, whereas other technical research communities (not limited to researchers in control theory) have embraced it to a much larger extent. This raises the question of whether there might be some sort of anti–fuzzy logic bias in the AI community, and if so where the roots of this bias might come from. I am by no means the first to raise this observation, but I've never been aware of any significant current discussion on this subject. Might this observation have merit or am I impugning a whole community unjustly?[30]

In this passage, the writer puts forward in the mode of testimony—based on his own personal experience and that of his colleagues—a more or less explicit antagonistic attitude toward fuzzy logic in the realm of "classical" artificial intelligence. The principal mediation that allows him to connect this antagonistic attitude with Elkan's de-monstration lies in the prize that the latter won at AAAI'93. For the author of this message, who makes himself the spokesperson for researchers in fuzzy logic, Elkan's de-monstration and the consecration it received clearly constitute the mark of a collective hostility toward fuzzy logic in artificial intelligence. Thus a conspiracy theory is suggested, along with the complicity of the AAAI prize committee, a group that supported and *relayed* Elkan's de-monstration, speaking for "artificial intelligence" as a whole.

At this stage I am not seeking to confirm or refute such propositions—or those concerning the more favorable "welcome" offered to fuzzy logic by the control systems community—by importing other data related to the question. In fact, it is a matter here above all of studying the way a new de-monstrative register, in comparison to those examined earlier, is introduced into the debates.

We observe de facto that Elkan's article is not simply subjected to critiques that use de-monstrative registers identical to the ones Elkan uses

[30] D. Lawler, message 853.

himself. In the last excerpt cited, the positing of a diffuse antagonism toward fuzzy logic leads to identifying the author's approach with a more general hostile attitude, beyond the precise analysis of the details of his de-monstration, which move to the background.

We see, then, that the de-monstrative registers of the original article, especially those used in the symbolic proof of the theorem, do not determine the ways in which the text is apprehended and "discussed," once it is in the hands of the commentators. We observe once again that Elkan's text can legitimately be compared to a user's manual. Although useful, such a manual generally proves to be insufficient to allow the user to reach the desired mode of operation. Even if Elkan's proof supplies a series of instructions for its own reading, it is nevertheless *read* in very different ways: it elicits responses in the form of electronic messages that the instructions alone do not succeed in determining in advance.

The list of de-monstrative registers deployed and associated with one another in this controversy thus keeps getting longer. Correlatively, it appears that the debates are shifting further and further away from a genre that we might have thought to be well defined at the outset.

However, while in the last excerpt studied Elkan's de-monstration is associated with a general antagonism toward fuzzy logic in the field of artificial intelligence, and is presented as a *mark* of that hostility or as its *result*, other modalities of association are deployed in other texts addressed to the forum. In the following excerpt, Elkan's original proof is on the contrary exhibited as a possible *cause* of such antagonism:

> There were several posts to comp.ai.fuzzy regarding the recent paper with the title "The Paradoxical Success of Fuzzy Logic" by Charles Elkan at the last AAAI conference (U.S. national Artificial Intelligence conference) this summer. The impact of the paper should absolutely not be underestimated. The AAAI proceedings are very well read in the AI community, and I fear that this article will destroy some of the trust that has so far been given to fuzzy logic in the AI community. This is a very dangerous situation as it might entail academic hostilities between the AI and FL communities. The paper is not badly written and even won a best-paper award at the conference. A refutation must be based on good arguments, and should be published in a respectable journal or better at a large conference. . . . Please refrain from flaming anybody, but good arguments are certainly welcome.[31]

Elkan's article is not presented here as the consequence of the existence of an antagonistic attitude toward fuzzy logic but as the possible *cause* of such a phenomenon, or of its exacerbation, given the visibility of the proceedings of the AAAI'93 conference. Once again, the details of the

---

[31] W. Slany, August 31, 1993, cited in message 1823.

proof are bracketed, and the debate shifts to an entirely different level. This time, Elkan's de-monstration is *assimilated* to a de-monstration of hostility and to an institutional risk for fuzzy logic.

To construct this register, the author adopts the posture of a sociologist of science. He presents researchers in fuzzy logic and in artificial intelligence as belonging to distinct communities. Without speaking of a preexisting cold war, he brandishes the threat of an "authentic" war. He sketches a sociological representation of the operation of science, which, as for Steven Shapin, attributes a central role to trust and credibility in the dynamics of recognition of scientific facts.[32] Notions of *respectability* and *trust* are thus mobilized to describe the ideal relations between the *communities* identified. In order to move toward that ideal, the author proposes actions envisaged in an institutional perspective, actions that highlight their own "respectable" character (the "respectable" form of texts, journals, and conferences in which the actions must be undertaken), relegating by the same token to the status of *formality* the very existence and expression of such counterproofs.

In the face of this type of de-monstrative register, certain participants, adopting an already well-identified approach, tried to bring back to light the absence of properties of fuzzy logic at the origin of these de-monstrations of "hostility." They sought to recenter the debates on the details of the symbolic de-monstrations. The attempt to displace the discussions that we have just brought to light was thus subjected to the trial of the production of de-monstrative registers that took the preceding messages into account, and that led to a broader but once again unitary debate.

Thus the author of the following text, who presents himself as a researcher in artificial intelligence, acknowledges the existence of hostility to fuzzy logic in his own field, even as he justifies this antagonism by the absence of the properties announced for that logic and other "methods":

> Of course there is anti-fuzzy, anti-Neuralnet, anti-GeneticAlg,[33] anti-HotNew-Method feeling in the AI community, but it is neither a bigoted nor a hidden bias. Underanalyzed, it may look like a "Not Invented Here" syndrome but it is really a matter of how results are presented and what claims are being made. ... The problem which hurts these subfields of AI is in the routine CLAIM, made implicitly or explicitly, that some magic exists in the WEAK METHOD (due to its intelligence, adaptiveness, biological plausibility, general-pur-

---

[32] Shapin, *A Social History of Truth.*

[33] The expression "genetic algorithms" refers to a method for the progressive development of systems inspired by a model from genetic biology. See Melanie Mitchell, *An Introduction to Genetic Algorithms* (Cambridge: MIT Press, 1996), and G.J.E. Rawlins, ed., *Foundations of Genetic Algorithms* (San Mateo, Calif.: Morgan Kaufmann, 1991).

poseness, or "strength") rather than in the HUMAN IN THE LOOP (who designed the task-appropriate knowledge structures). . . . —Jordan Pollack, Assistant Professor CIS Dept/OSU Laboratory for AI Research.[34]

Jordan Pollack refuses to let the debate over the absence of certain properties of fuzzy logic be short-circuited by considerations having to do with the existence of hostility to fuzzy logic in the artificial intelligence community, for, according to him, the first phenomenon is the cause of the second. Fuzzy logic, like neural networks and genetic algorithms, is for the occasion constituted as a subfield of artificial intelligence. Pollack tries to bring to the fore as a *justified* and *declared* antagonism, based precisely on the absence of strong properties in the incriminated "methods," what some perceive as a *dissimulated* patriotic reaction that is unfavorable to those subfields (presented implicitly as developed outside of the United States).

The denunciation of the fraudulent attribution of the operation of devices described as fuzzy to the properties of fuzzy logic takes a new turn here, however. The author presents the creators of such devices as the "cause" of their successful operation; the "methods" in question are thereby stripped of their virtues. Pollack's de-monstrative strategy here is quite distinct from Elkan's, since the negation of the role of fuzzy logic in the successful operation of putative fuzzy systems does not proceed from the advancing of formal inconsistencies but from the exclusive invocation of virtues proper to the skills of the individuals who have perfected the machines.

However, the string of viewpoints does not end here. The exchanges in the forum give rise to other attempts to neutralize the debates over the details of the proofs and correlatively to other displacements of attention, to other visions. Certain messages do not identify or associate Elkan's demonstration so abruptly with the existence of a general antagonism toward fuzzy logic. They stress the mediations of this antagonism by explicitly denouncing the *complicity* and the disparaging attitude on the part of those who reviewed Elkan's article and of the members of the AAAI'93 prize committee. An excerpt from a message by E. Ruspini offers a good illustration of this:

> We [people of the fuzzy logic community] were both amused and perplexed as to how Elkan's paper (ignoring results so well known that they appear in textbooks) not only passed the referees but actually won an award. Normally, we would not bother to clarify a simple matter that is otherwise obvious to anybody that actually bothers to learn about FL before trying to criticize it (FL skeptics keep popping up repeating worn arguments that were disqualified long

[34] J. Pollack, message 856.

ago—we are rather busy doing positive things and cannot answer to all of them). The notoriety that NCAI [National Conference on Artificial Intelligence] has given to this work requires, on the other hand, some response. . . . This work has received particular attention from non-specialists since it was the only paper on fuzzy logic presented at the 1993 National Conference on Artificial Intelligence, being also the recipient of one of the "best paper" awards. . . . One can only wonder to what kind of analysis, if any, this claim was subjected by the NCAI referees, when the author claims, in the discussion following his theorem, that "What all formal fuzzy logics have in common is that they reject at least one classical tautology, namely, the law of the excluded middle," while basing his main argument on an equivalent assertion. . . . Much has been said recently about overzealous hype among those propounding fuzzy logic. While quite a few of these claims are certainly true, it is clear from papers such as Elkan's that not enough is being done in certain forums to assure that papers dealing with fuzzy logic, either pro or con, are subjected to a fair and competent review process. Many technical conferences and meetings are, in the views of many in the fuzzy-logic community, unfairly hostile to fuzzy logic, while being, on the other hand, ready to accept the work of skeptics with nary an effort to determine its value. Elkan's paper should have never survived the refereeing process, let alone be awarded a prize.[35]

Ruspini's intervention constitutes, performatively, a "community" of researchers in fuzzy logic whose spokesperson is the author himself. The critiques formulated in this message about authors whose identity is not revealed underline the divisions of this "community" at the outset. This time, Ruspini is openly denouncing the role of the reviewers and the AAAI'93 prize committee members, along with their disparaging attitude toward fuzzy logic. The specificity of this register lies not only in the targeted critique of the conference organizers, but also in the way the mediations of the hostility evoked are articulated.

Ruspini thus refers to the existence of an opposition to fuzzy logic that predates the drafting of Elkan's proof, even as he presents the acceptance of Elkan's article and the prize it won as a vector of that antagonism. In other words, his de-monstration carries out a synthesis of the preceding positions, presenting Elkan's article as both consequence and cause of an unfavorable judgment with respect toward research carried out in fuzzy logic.

This de-monstrative register takes a particular stand against the view according to which the correctness of Elkan's proof is attested by the trial of the article-selection process and by the prize that Elkan's contribution won at AAAI'93. For this view, Ruspini substitutes another by stressing

[35] E. Ruspini, message 816.

the recourse to unfair behavior: Elkan's de-monstration becomes the expression of a collective act of ill will and a sort of local conspiracy.

Moreover, this denunciation is peculiar in that it is not carried out in an exclusive fashion, since it accompanies a critical discussion (one we have already examined) focused on the details of the proofs. The proposed new representation of Elkan's de-monstration is thus also based on a *transposition* of the resources associated with the original proof. In the context of the excerpt under discussion, furthermore, Ruspini insists on the fact that Elkan mobilizes the principle of the excluded middle in his proof, even though Elkan has presented that principle as rejected by fuzzy logic.

As this proliferation of registers continues, we observe once again the recourse to a discourse on self-evidence, which takes an unprecedented turn here. By adopting a professorial tone and by referring to fuzzy logic textbooks, Ruspini helps make Elkan look like a "dunce" who has not taken note of the most "immediate" scholarly evidence. The discourse on self-evidence thus takes the form of an argument from a position of authority, based on the institution of a teacher-student relationship.[36]

Implemented in Elkan's case, the inauguration of a teacher-student relationship is then extended to other situations. Ruspini evokes precedents, comparable breaches with regard to self-evidence. What is at issue here is not just bias, but also a form of *incompetence* that is represented as widespread among researchers in artificial intelligence.

On this occasion, we note once again an important dimension of the de-monstrative activity: the protagonists in the debates are taking into account and actively managing the de-monstrations' visibility. Ruspini evokes the economy of his own de-monstrative actions, as did another contributor who, in a previously analyzed message, called for the formulation of *respectable* de-monstrations to defend the credibility of fuzzy logic, in the face of the *risk* represented by the visibility of Elkan's article. Continuing to use a self-assured if not condescending tone (a de-monstrative strategy in its own right in the jousting matches we are observing), Ruspini thus asserts that he is deigning to respond to Elkan's article not because his critique would be of any particular value but because of the alleged visibility of Elkan's text.

In the last analysis, we see that Ruspini's message accumulates a large number of registers destined to shore one another up, and to support

---

[36] Here again we find a type of argument frequently used in the process of teaching logic described above. The claim that a given manipulation was self-evident constituted a resource often used by the instructors in their attempts to put an end to the production of student questions, deemed overly inflationist, or to end a debate with students when the outcome was uncertain.

in particular the thesis of the existence of a pronounced and sustained antagonism to fuzzy logic in artificial intelligence. In the string of demonstrations by the forum's participants, this last viewpoint nevertheless runs up against messages conveying competing representations, on the basis of personal *testimony.*

Thus we recall first of all that Elkan denied being inspired by any desire to do harm. In particular, he asserted that it was not at all his intention to call back into question the "usefulness," the proper functioning, and the "success" of fuzzy systems.[37] Other testimony tends in the same direction, as for example the following:

> I have nothing against fuzzy logic—it has proven itself useful in many (quite diverse) practical situations, but I do object to the characterization . . . that all work done without fuzzy logic up to this point is somehow invalid. I'm all for fuzzy logic, where it suits the job (the whole idea is quite fascinating to me), but get off your fuzzy high horse.[38]

In this message, as in Elkan's, the author grants fuzzy logic a certain value, however limited and essentially practical it may be. Still, he does not declare himself ready to accept theses that emphasize more general virtues. His tone, like Elkan's, suggests the adoption of a benevolent and impartial attitude, one that contradicts the preceding assertions. The de-monstrative register at stake proceeds from a reversal of perspective, since, conversely, the author stigmatizes the disparaging attitudes of fuzzy logic proponents toward artificial intelligence. A new representation (or, rather, a new transposition) of Elkan's de-monstration results; it is suggested that Elkan has adopted the proper tone in response to the condemnation by researchers in fuzzy logic of the work done in artificial intelligence.

As we see, the stagings of antagonisms destined to neutralize or shore up the discussions of the details of the formal de-monstrations vary considerably from one message to another. More particularly, we may recall that certain authors attributed the existence of hostility to fuzzy logic to "closed" minds[39] and that one contributor denied that researchers working on artificial intelligence in the United States had adopted a patriotic attitude while rejecting fuzzy logic.[40] In fact, the accusations of closed-mindedness, associated with charges of irrationality or denunciations of conservatism, themselves appear in the most diverse configurations. The following supplementary declensions illustrate the point perfectly:

[37] C. Elkan, message 794.
[38] W. Dwinnell, message 769.
[39] See once again J. Wiegand, message 767.
[40] J. Pollack, message 856.

His point seems to be, "it's not really Engineering." This is an unfortunate viewpoint that highlights the conservative nature of engineering. Elkan is trying to stop a paradigm shift.[41]

Well, it is nice to see the fuzzy community approach such a serious challenge with an open mind. . . . I did not reply to his [Elkan's] post since it was directed at an emotional level.[42]

The author of the first message challenges Elkan's de-monstration without looking into its details. He assimilates it to the expression of a conservative reaction he deems characteristic of the realm of engineering science. The staging of such conservatism is shored up by a Kuhnian tone,[43] making fuzzy logic a new paradigm. The presumed hostility to fuzzy logic appears here as a form of "resistance to progress."

The second message mobilizes still another de-monstrative register. The attempt to neutralize Elkan's de-monstration proceeds from a denunciation of an irrational attitude on the part of the representative of researchers in artificial intelligence, an attitude that contrasts with the rational behavior and "open-mindedness" attributed to the researchers in fuzzy logic.

Thus we find once again in the forum situations of a sort brought to light by Erving Goffman.[44] We are dealing with actors who seek to institute different modes of interaction in confrontations with one another; this leads to conflicts, negotiations, and opposing normative positions on ways of conducting a dialogue.[45] The conflicts and negotiations bearing on the practices of de-monstration that we have previously observed finally represent only one particular aspect of the struggles over the modes of intervention in the forum.

However, it is appropriate to stress the fact that the modalities of this double negotiation, which bears *simultaneously* on the de-monstrations and on the ways of debating them, constitute so many possible ways of supporting or neutralizing the positions defended in the context of the discussions of the details of the proofs. The preceding excerpt offers a good example: Elkan's de-monstration appears to be neutralized by the evocation of the inadequacy of its author's mode of intervention.

---

[41] J. Wiegand, message 767.

[42] J. Wiegand, message 832.

[43] See Kuhn, *Structure of Scientific Revolutions*.

[44] Goffman, *Presentation of Self*.

[45] On this topic, see also Steven Shapin's analyses of the role played by the institution of rules of politeness in the dynamics of recognition of scientific facts, in *A Social History of Truth*.

But let us come back to the question of the deployment of demonstrative registers proceeding from denunciation of closed-mindedness. If the two preceding texts supply good illustrations of this, we must nevertheless note that these registers can take the form of a denunciation based on a culturalist position. In this circumstance, Elkan's adversaries combat his de-monstration by identifying "traditional" artificial intelligence—linked to a "limited" Western mode of thought—with a source of oppression of fuzzy logic and a "corresponding" Eastern mode of thought.

CONTESTING A PROOF AND DEFENDING LOGICAL PROPERTIES BY EVOKING A CULTURAL SPECIFICITY

Certain messages attempt to defuse Elkan's critique and in particular to show that fuzzy logic is not the victim of the inconsistencies evoked, by positing a cultural incommensurability between the approach taken in Elkan's proof and "the" approach attributed to fuzzy logic. In these texts, fuzzy logic is presented as indissociably linked to the Eastern mode of thought and to Asian civilization. The concepts of binary logic, especially "the" (Western) notion of paradox, associated with Western civilization and a tradition of Aristotelian thought, would not make it possible to grasp the subtleties of fuzzy logic.

By adopting this culturalist position, certain authors attempt to show that the inconsistencies Elkan evokes are *relative* to a mode of thought that valorizes a particular notion of precision, one that cannot be retained in efforts to conceive of the coherence that is proper to fuzzy logic and the value of fuzziness.[46] A new de-monstrative register is thus introduced to short-circuit the debates over the details of the proofs. This register proceeds from a form of *cultural relativism*.[47] The following excerpt illustrates this approach:

>Last week at AAAI, a paper by Charles Elkan of UCSD was given
>with the above title. It concludes that a "standard version" of fuzzy
>logic collapses mathematically to binary logic.

[46] These debates highlight the interest of developing a social history of the categories of rationality. The celebration of fuzziness rather than precision, as an attribute of rationality, no doubt constitutes an object for historical analysis in its own right. By comparison, see the categories consecrating the rationality/irrationality dichotomy (such as, here, the precision/fuzziness dichotomy) analyzed in Dominique Pestre, *Physique et physiciens en France, 1918–1946* (Paris: Archives contemporaines, 1984), pp. 171–207.

[47] It should be noted that this form of slippage from cultural relativism to logical relativism can be compared to the possible passage from cultural relativism to aesthetic relativism. On this subject, see Pierre-Michel Menger, *Les laboratoires de la création musicale: Acteurs, organisations et politique de la recherche musicale* (Paris: La Documentation française, 1989).

I should hope so! Fuzzy logic is the generalization of binary logic. It sounds as if Mr. Elkan has slipped back into the trap of Aristotelian logic. Witness the title: "The Paradoxical Success . . ." There are no such things as paradoxes, only semantic dead-ends that result from the Western mindset.[48]

The de-monstrative schema used in this message is easy to identify. Rejecting the very notion of paradox, the author opposes the patient de-monstrative efforts used to exhibit inconsistencies in fuzzy logic. The notion of paradox, mobilized by Elkan to stigmatize an impasse proper to "the" *axiomatics* of fuzzy logic, is exhibited here as an impasse inherent in the Western *mode of thought*. To organize this shift of focus, the writer identifies and then disparages as a whole "the" tradition of thought of "Western civilization in general." The writer is not simply adhering to cultural relativism here; he moves on to establish a hierarchy among the cultures he identifies. The exhibition of fuzzy logic as a generalization of binary logic (also labeled Aristotelian) leads him to consecrate an Eastern mode of thought as triumphing over the failures (the "impasses") of the Western mode of thought. Asserting the opposite of the text he is citing, the writer thus offers a new representation of Elkan's de-monstration, which is reduced to an absolute *misunderstanding* of fuzzy logic that originates in a fundamental cultural *limitation*.

In the face of such a positioning of the debate, several participants intervene to denounce an excessive culturalism. We can analyze the various de-monstrative strategies used to confront this new register by studying excerpts from a few messages. Let us begin by examining a response formulated by Elkan himself to this type of critique:

> PHILOSOPHICAL AND MATHEMATICAL ISSUES The claimed binary opposition between a Western discrete mindset and a continuous Eastern mindset is ridiculous. In the history of Western thought, there have always been competing holistic, continuous and reductionist, discrete points of view. It makes reasonable sense to talk of "Western" thought because there has been continuity in philosophy from the Greeks to the present. I know almost nothing about non-Western philosophy. However I am willing to conjecture that "Eastern" thought is much less of a unitary tradition, and that in any particular division of Eastern thought (for example Confucianism) there are also parallel strands of continuous and discrete points of view.[49]

We have already seen the way notions of paradox or inconsistency could be rejected in the name of a cultural incommensurability between Eastern and Western modes of thought. To defend his proof and bring back into view inconsistencies proper to fuzzy logic, Elkan adopts a de-

[48] J. Wiegand, message 767.
[49] C. Elkan, message 794

monstrative strategy consisting of calling the nature and scope of this dichotomy back into question. To this end, he sets himself up as a historian of civilizations and philosophy: he challenges the distinctions made between two modes of thought by evoking the diversity of their respective tendencies, their hybridization, even their possible unity; in the process, he formulates a thesis about the continuity of Western philosophical thought from the Greeks to our own day, and develops hypotheses on the history of the various tendencies of Confucianism.

We have already seen how debates about logic can pass through the elaboration and discussion of theses about the philosophy of logic; here we see to what extent this activity can also consist in writing or rewriting the cultural history of logic, philosophy, and civilizations. By this very token, we note to what extent it would have been mistaken to think that the debates over a logical theorem such as Elkan's belong to a genre that is well defined in advance, in which only symbolic proofs are introduced.

To be sure, we might not have taken these excerpts into account; we might have taken considerations about the details of the proofs as our sole object. But we would then have been very far from grasping the dynamics of the debates. In particular, we would not have perceived the way in which the formulation of theses about the history of philosophy and Confucianism could constitute an important element for defending the correctness of a proof. Correlatively, we would not have been in a position to grasp all the aspects of the work of the actors, which, far from being limited to the drafting of symbolic proofs, also consists in producing reflections stemming from epistemology, from the history of science, or even from philosophy and history in general.

In fact, we note that the development of the exchanges on comp.ai .fuzzy is related to the introduction of an increasing volume of considerations, including competing viewpoints as to what it is acceptable and important to debate, and what can legitimately stem from the exercise of logic. In other words, the disagreements spread to the definition of borders between what may or may not constitute an acceptable element for discussion, between what may or may not be considered as legitimately arising from a debate in logic, or between what does or does not constitute a "strong" argument. Since these borders are not clear for the group, and since their negotiation represents springboards for advancing the "discussion," we understand better that it would hardly have been possible, given our problematics, to study only some of the elements of the debates while skipping others.

To return to our initial question, how then does Elkan manage to face up to this deployment of heterogeneous de-monstrative registers, often different from those on which he himself drew in his article? He appears unquestionably endowed with a powerful capacity for adapting in order to defend his article, to become its spokesperson, by producing new de-monstrations. He displays a pronounced aptitude for integrating the multiple registers of his interlocutors into his own formulations. He manages to place himself on their terrain in order to constitute—or to shatter, as the case may be—relationships among the most varied arguments. This specific capacity is clearly inscribed in the extension of his training and his career as a researcher in computer science, which gives him the ability to dialogue with actors working in a large number of fields; we shall look more closely into this aspect later on.

However, Elkan also benefits from the support of converging points of view. Thus several participants express decided and sometimes virulent opposition to the type of culturalist thesis examined above, as the next two excerpts attest:

> I do object to the characterization of anyone who doesn't use or accept fuzzy logic as "Aristotelean" or "Western," as if fuzzy logic were the Holy Grail, and the entire universe has changed since its introduction.[50]

> Dennis Francis Hefferman (Yes, the address is spelled wrong and no, it's not my fault.) Montclair State College. Comp Sci/Philosophy Taoism in two words: CHILL OUT![51]

The attempt to neutralize Elkan's de-monstration by recourse to culturalism hardly garners the support of all. It is met with vigorous hostility from contributors who deem this register unacceptable. In the first message above, the writer thus contests a form of reductionism that consists in judging that the rejection of fuzzy logic by an individual would stem essentially from the latter's Western origins. He also simultaneously opposes any reconstruction of the history of civilizations around the "discovery" of fuzzy logic, targeting the praise of fuzzy logic and Eastern civilization simultaneously.

In the second message, the rejection of culturalist positions takes a more ironic form. While a number of the protagonists put little formulas intended to be humorous tags at the end of their messages, in Dennis Francis Hefferman's case this takes the form of a mocking critique of an "ideology" whose victims include researchers in computer science and philosophers.

The shared rejection of this cultural relativism paired with a form of *Orientalism* thus gives Elkan an opportunity to rally actors around his cause. Antagonism constitutes the glue binding together several contributions intended to neutralize a de-monstrative register that threatened El-

---

[50] W. Dwinnell, message 769.
[51] D. Hefferman, message 834.

kan's de-monstration, with at least indirect support for the latter as a consequence. On this basis, one might find it surprising a priori that certain researchers in fuzzy logic subscribe to such a denunciation. Such is nevertheless the case, as the following excerpt from one of Ruspini's messages shows:

> In particular, I pledge to stay away from claims based on the notion that the purported paradox requires explanations based on esoteric mysticisms. . . . I am bemused, on the other hand, by the efforts of some who have recently posted messages in bulletin boards, trying to explain Elkan's errors on the basis of arguments rooted on Eastern mysticism and on the limitations of the "Western mindset," following a recent trend initiated by questionable articles and books about fuzzy logic written by "philosophers" and "experts."[52]

Ruspini, as we have already seen, develops multiple criticisms of Elkan's de-monstration; still, he refuses to use culturalism as an argument. To defend his proof in the face of cultural relativism, Elkan benefits from a singular conjunction of interests based on the divisions that affect the community of researchers in fuzzy logic, and on disagreements over the nature of their research objects. Ruspini refers explicitly to these divisions, helping by that very token to reinforce them, even if he names no names—we shall have the occasion to look more closely at these conflicts later on.

If we judge solely by the hostile reactions to this register, it seems likely that Ruspini thought that the Orientalist positions constituted a de-monstrative resource that was more harmful than beneficial to the image of fuzzy logic. According to this hypothesis, denouncing this register amounted for his part to attempting to avoid the discreditation of his own research objects. On more than one account, the collaboration of a researcher in fuzzy logic such as Ruspini in the neutralization of a register that was "devastating" for Elkan's demonstration is thus hardly surprising.

Moreover, we can also see in it, as in other messages, an implicit condemnation of the "politically correct." It is indeed highly likely that the de-monstrative registers based on cultural relativism and Orientalism were perceived by a large number of their detractors as stemming from a stance of political correctness. As evidence, we may consider the association, this time made overtly, in the following excerpt:

> \>It sounds as if Mr. Elkan has slipped back into the
> \>trap of Aristotelian logic. Witness the title: "The
> \>Paradoxical Success . . ." There are no such things
> \>as paradoxes, only semantic dead-ends that result
> \>from the Western mindset.

[52] E. Ruspini, message 816.

I'd like to see such a remark in a proposal or the preliminary remarks in a system or software design document. Somehow, I don't expect to. Fuzzy seems to attract rhetoric (here PC?) or some reason. I would be interested to know if it also does in Japan. One possible reason for Fuzzy's popularity there is that it fits iterative development so well[53]—not because of some attraction for imprecision. Controlling subway cars doesn't have much in common with Zen archery.[54]

In this message, the writer stigmatizes, cautiously but explicitly, the politically correct discourse in the text he is citing (Wiegand, message 767). Once again, Elkan benefited from support based on the rejection of culturalist positions. Making fun of the register of political correctness, the writer organizes the shift of focus toward the objects of earlier debates. He exhibits a property of fuzzy logic as a substitute for the culturalist theses.

To combat the Orientalist positions, the writer adopts the following de-monstrative strategy: he puts forward a *universal* property of fuzzy systems and their correspondence with Japanese working methods, in order to account for their defense in the Land of the Rising Sun. In other words, he exhibits a *discourse of celebration*, and reconstructs its situated emergence by invoking one specific factor: a structural identity between the techniques for perfecting fuzzy systems and local working methods of the iterative type.

This approach is in fact very close to Elkan's. Let us recall indeed that the latter, in an excerpt analyzed above,[55] called back into question the explanatory character of a dichotomy operated between Eastern and Western modes of thought, while drawing his readers' attention to other causes of the singular development of fuzzy systems in Japan. On that occasion, he invoked a set of *socio-techno-economic factors*: the state of the market in high-technology goods and the correlation between methods for perfecting fuzzy systems with the specific mode of organization of labor, production, and management of Japanese businesses.

Several messages addressed to the forum proceed in fact from a similar de-monstrative strategy, in several declensions. Among the latter, let us note the various ways in which a matter of fashion in Japan is invoked: an *infatuation* with fuzzy products on the national scale would explain

[53] In using the expression "iterative development," the writer, like Elkan in a message we have already seen (message 794; see above, note 6), is probably referring again to a working method that is often perceived as "typically Japanese," consisting, for example, when it is a matter of perfecting controllers, in gradually improving their programming by adjusting different numerical coefficients or different operating rules, one step at a time.

[54] H. Bonney, message 787.

[55] C. Elkan, message 794.

the surge in their development. This thesis is suggested in the following passage:

> >One possible reason for Fuzzy's popularity there is
> >that it fits iterative development so well—not
> >because of some attraction for imprecision.
> >Controlling subway cars doesn't have much in common
> >with Zen archery.
> Apparently not, it seems that in Japan, Fuzzy Set Theory is a fad and anything fuzzy is to the consumer a warm and . . . (wap!) . . . {sorry}[56]

This contributor rejects the culturalist arguments in his turn even as he gives a new orientation to the readers' shift of focus, a shift initiated by the author he cites. Instead of exhibiting a discourse celebrating fuzziness and explaining its emergence by invoking a structural equivalence between techniques for perfecting fuzzy systems and traditional working methods, he refers to a "social" phenomenon, a "fashion" affecting Japanese consumers. The Japanese are not shown here as determined in their behavior by a distinctive "mode of thought." Their singular behavior is explained as a function of a local phenomenon, whose principle ("fashion") is not itself associated with any Eastern cultural specificity.

Here, the indirect defense of Elkan's de-monstration in the face of Orientalism thus passes through a *reformatting* of the Japanese "mode of thought," grasped according to a more universalist mode. This reformatting is pursued in other texts, moreover, through the invocation of linguistic particularities. In these messages, instead of invoking differences in mental structures, the writers put forward specific connotations of the term "fuzzy" in English and in Japanese in order to account for the differentiated "infatuations" with fuzzy logic in the United States and in Japan. The following messages are illustrative:

Fuzzy seems a very practical tool for us. Now if it only had a better name . . .[57]

> >The fact that Japan leads the world in fuzzy patents
> >and in fuzzy controllers has several plausible non-
> >mystical explanations: . . .
> Here is another: Both Japan and America use the same word "fuzzy." In America, the word has many old negative connotations. In Japan, it has no old connotation, positive or negative. (And it is a wonderful sounding word.) Thus, "fuzzy camera" sounds high-tech in Japan, and goofy in the U.S.[58]

[56] M. Aichlmayer, message 789.
[57] H. Bonney, message 787.
[58] C. Kadie, message 798.

In these two texts, and more specifically in the second, particular attention is paid to the word "fuzzy" in order to account both for a certain hostility to fuzzy logic in the United States and for the attractiveness of fuzzy logic in Japan. The writers do not invoke the existence of distinct mental structures that would characterize Japanese people on the one hand, Americans on the other. They highlight the dissimilar *effects* that the use of the same term in two different spaces allegedly induces, as a function of linguistic idiosyncrasies. In other words, the reformatting of the Japanese "mode of thought" that we are considering here depends on the substitution of a *sociolinguistic* register for an Orientalist register. It is no longer a question, in these messages, of a gap between two civilizations, but simply of *language effects*. The people concerned are no longer so foreign; they are endowed with identical cognitive and behavioral faculties.

In these two messages and the preceding ones, then, we see how the contestation of a proof or its defense, even indirect, can *operate by way of* the deployment of de-monstrative registers having to do with the nature of the modes of thought of human beings.[59] Comparatively, the pro- or anticulturalist approaches developed by certain sociologists of knowledge seem susceptible to analogous treatment: they can themselves be taken not as resources for analysis but as objects of study.

The study of the corresponding literature in fact shows that the oppositions between authors stem from the use of resources similar to those mobilized by the actors whose messages we have just examined. In the history of the social sciences, the discussion of the positions proper to cultural relativism appears in part to be correlated with the proffering by certain analysts of the incommensurable character of the forms of logic used by various human societies. For example, Lévy-Bruhl assigned prelogical mentalities to certain societies, societies that, according to him, did not systematically respect "the *laws*" of logic.[60] Granet opposed a Western mode of thinking through syllogisms to the Chinese mode of thinking through analogies.[61] Peter Winch did not affirm that Azande thought was characterized by the same logic as that of Westerners.[62]

[59] Such a situation hardly constitutes a unique case in the history of logic. For an initial glimpse of the abundant disagreements and controversies over the "nature of the human mind" to which the proffering of competing definitions and practices of logic has led since the early twentieth century (in particular in Church, Gödel, Husserl, and Hilbert), see Largeault, *La logique*, pp. 66–69, 112.

[60] Lévy-Bruhl, *Primitive Mentality*.

[61] Granet, *Études sociologiques sur la Chine*.

[62] Peter Winch, "Understanding a Primitive Society," *American Philosophical Quarterly* 1 (1964): 307–24.

Conversely, the production of de-monstrations intended to combat the theses deemed culturalist, through the mobilization of various registers (socio-techno-economic or others), is also found for example in Bloor, when he refuses to attribute systematic nonrespect of the principle of contradiction and a distinct form of logic to the Azande.[63] Starting from his reading of Evans-Pritchard,[64] Bloor causes institutionalized "informal arguments" to *appear*, arguments that circumvent "formal deductions" compatible with the principle of contradiction.[65]

Among the contributors to the comp.ai.fuzzy discussions, as among the authors mentioned above, the production of pro- or anticulturalist de-monstrations is an interesting object of study to the extent that one can see how discourses about mental structures, cultures, and logic are constituted simultaneously. In one and the same movement, these actors are led to negotiate the definitions of logic or logics (constituted or not for example around some "immutable" principles, such as the principle of contradiction), cultures, or the cognitive capacities of individuals. In other words, the analytic posture that we are evoking is distinctive in that it aims to account for the simultaneous *production* of certified logical, sociological, and psychological knowledge, instead of starting from potentially *static* and *definitive* descriptions of logic or logics, cultures, and human cognitive capacities.[66]

Similarly, a political denunciation of the oppression of minorities by a given form of logic, instead of serving the principle of analysis as it does for Andrea Nye,[67] can be usefully considered as an object of analysis. Studying the mobilization or the denunciation of political correctness makes it possible here to get a better grasp of the dynamics of validation of a proof and the modalities of negotiation of the normative definitions of a logical theory (fuzzy logic), by bringing to light an important mediation of the debates.[68]

---

[63] Bloor, *Knowledge and Social Imagery*, pp. 142–43.

[64] Evans-Pritchard, *Witchcraft, Oracles, and Magic*.

[65] See also Geoffrey Lloyd's critique of the uses of the notion of mentality, in G.E.R. Lloyd, *Demystifying Mentalities* (Cambridge: Cambridge University Press, 1990); and see Raymond Boudon, *The Art of Self-Persuasion: The Social Explanation of False Beliefs*, trans. Malcolm Slater (Cambridge, Mass.: Polity Press, 1990).

[66] See Latour and Woolgar, *La vie de laboratoire*, pp. 10–11. In this text, Latour recalls the alacrity with which anthropologists working with ORSTOM (now IRD, a research institute focusing on development) were prepared to explain the difficulty French companies had finding competent Ivoirian managers by invoking a cognitive specificity, speaking of an "African mentality, the Black soul, and psychology." See also Bruno Latour, "Les idéologies de la compétence en milieu industriel à Abidjan," *Cahiers Orstom—Sciences humaines* 9 (1973): 1–174.

[67] Nye, *Words of Power*.

[68] It would be a mistake to believe that the deployment of political considerations associated with the development of logical formalisms is a singular phenomenon, one that would

Finally, an important result achieved by the approach adopted in this chapter certainly lies in the fact that it has brought to light the considerable diversity of the de-monstrative registers deployed in order to defend or invalidate Elkan's de-monstration or its substitutes. To grasp the dynamics of these debates, we have not limited our examination to symbolic proofs alone. We have had to follow a controversy that was developing in multiple directions. We have thus observed that a great variety of objects, black boxes, and procedures for *attributing* logical properties was associated with the collective statement "fuzzy logic." As surprising as it may seem, the appreciation of Elkan's de-monstration and its substitutes, and, correlatively, that of fuzzy logic, has been seen to pass through a large number of heterogeneous trials.

A list of these trials would include the display of devices for de-monstrating the properties of fuzzy logic (fuzzy trains, elevators, or helicopters), paternity trials, the highlighting of expertise bearing on the organization of labor and business management practices in Japan, the deployment of socio-techno-economic mediations, appeals to self-evidence, the establishment of relationships between registers perceived (or not) as heterogeneous, the estimation of the intensity of the dialogue between the protagonists in the debates, the assimilation of Elkan's de-monstration to a general hostility to fuzzy logic in the artificial intelligence community, the attribution of the successful operation of fuzzy systems to the skills of their inventors, the denunciation of a plot fomented by the organizers of a conference, the stigmatization of incompetence and malicious actions, denunciations of closed-mindedness and irrational or conservative behavior, normative considerations regarding the legitimate modes of participating in debates, the invocation of Confucianist, Taoist, or Zen traditions in opposition to an Aristotelian tradition, the rewriting of the history of Western philosophy and of Confucianism, the recourse to positions stemming from cultural relativism in order to reject the very notion of paradox, the hostility to Orientalism and to the politically correct, the formatting and reformatting of the Japanese mode of thought.

---

depend on the specificity of fuzzy logic. On the contrary, the history of logic is full of comparable cases. Thus, for example, we note that a number of texts written in Europe during or immediately after World War I or World War II associate the development of logic with the return to peace among and within nations. See for example the lecture given by Hilbert in Zurich on September 17, 1917, during the annual meeting of the Swiss Society of Mathematicians (David Hilbert, "Axiomatisches Denken," *Mathematische Annalen* 78 [1918]: 405–15). In the aftermath of World War II, see in particular Marcel Boll and Jacques Reinhart, *Histoire de la logique* (Paris: Presses Universitaires de France, 1946), pp. 9, 56. See also the way logic is presented as a bulwark capable of preserving the "individual" against ideologies, in Merrie Bergmann, James Moor, and Jack Nelson, *The Logic Book* (New York: Random House, 1980), pp. 1–2 (this is a textbook used in the United States to teach logic at the university level in the early 1990s; for more details on this, see Rosental, "De la logique au pathologique").

It must be noted that these diverse ways of contesting Elkan's de-monstration and its substitutes were the object of variable combinations. Each participant in fact produced only one component, and defined those that struck him as legitimate or on the contrary unacceptable. The corresponding subsets varied from one actor to another, a fact that, as we have seen, constituted an object of polemics in itself.

The introduction and association of diverse de-monstrative registers in the debates came about with great fluidity. In the final analysis, this fluidity was equaled only by the alacrity with which the participants challenged the procedures for attributing logical properties to technological devices. The inflation in the production of de-monstrative registers was thus the correlative of the rarity of convergence around the viewpoints of the various writers, as we had already seen in connection with the mere deployment of symbolic proofs.

But the proliferation of de-monstrative registers also corresponded to the diversity of the involvement of logical objects in multiple technological fields. We have once again encountered the "impure" objects—science, technology, and society mixed together—with which sociological investigations into the experimental sciences must deal. For the considerable number of technological devices whose operation was attributed to fuzzy logic helped make fuzzy systems vectors just as essential to the display of the properties of fuzzy logic and to the discussion of the correctness of Elkan's theorem as the symbolic proofs were.

Clearly, if we want to understand what happened to Elkan's theorem in the wake of its publication, we cannot limit ourselves to an analysis of the debates that took place on comp.ai.fuzzy. We have to study the extent to which these exchanges were inscribed within a broader dynamics, going beyond the confines of the electronic forum. We have already seen that the contributions to this forum could not be reduced to exchanges of *arguments*;[69] we can now stress the limits of this model even more clearly by broadening the scope of our investigations.

In order to gain a better grasp of the dynamics of production of de-monstrations (in particular Elkan's), we need to pursue our observations in other spaces; we need to study to what extent this specific work may be accompanied by other endeavors. It is a matter of circumscribing as much as possible the details and intricacies of these latter, and of understanding how they may be inserted into individual or collective strategies that are deployed over the long term.

[69] Let us recall that this point constitutes an advance in relation to the type of analysis proposed by Lakatos for the controversy over Euler's theorem. Owing to the procedure of narrative stylization Lakatos adopted, his analysis was actually reduced to a study of argumentation. See Lakatos, *Proofs and Refutations*.

In other words, it is important to apprehend the specific roles of the preceding messages more precisely, by situating them in relation to dynamics that may come into play outside the electronic forum. In examining the way actors manage the visibility of the de-monstrations, we have already begun to see that the work of de-monstration cannot be summed up as a fixed and unproblematic confrontation between a text and its readers. This work also relies on a great number of mediations (among others, various actions of *accompaniment*) that play an essential role in the *evolution of the actors' positions*; these remain to be examined in detail.

Moreover, while we have seen that an unequal visibility of the various de-monstrations was organized by certain actors, it is important to ask whether key positions do not finally emerge from the debates. It is a matter thus not only of trying to better apprehend the dynamics of production of the messages, but also, since we are interested in the potential collective validation of Elkan's theorem, a matter of investigating the developments of the controversy that came after the exchanges on the forum. Of course, given our problematics, it will also be important to carry out investigations into the period prior to the publication of Elkan's theorem in order to try to grasp its emergence in the wake of its earliest drafts. But let us begin by looking at the actions carried out by the actors in order to *accompany* their production.

PART THREE

MEDIATIONS USED TO ADVANCE

A LOGICAL THEOREM

"Wouldn't it be possible for me to give a party, for people to hear your
friend play?" murmured Mme. de Mortemart, who, while addressing herself exclusively to M. de Charlus, could not refrain, as though under a
fascination, from casting a glance at Mme. de Valcourt (the rejected) in
order to make certain that the other was too far away to hear her. "No she
cannot possibly hear what I am saying," Mme. de Mortemart concluded
inwardly, reassured by her own glance which as a matter of fact had had a
totally different effect upon Mme. de Valcourt from that intended: "Why,"
Mme. de Valcourt had said to herself when she caught this glance, "Marie-
Thérèse is planning something with Palamède which I am not to be told."
. . . Then without paying the slightest attention to her silent prayers, as she
made a smiling apology: "Why, yes . . ." he said in a loud tone, audible
throughout the room. . . . Mme. de Mortemart told herself that the aside,
the pianissimo of her question had been a waste of trouble, after the mega-
phone through which the answer had issued. She was mistaken. Mme. de
Valcourt heard nothing, for the simple reason that she did not understand
a single word. . . . She [Mme. de Mortemart] intended, on the morning
after the party, to write her one of those letters, the complement of the
revealing glance, letters which people suppose to be subtle and which are
tantamount to a full and signed confession. For instance: "Dear Edith, I
am so sorry about you, I did not really expect you last night" ("How could
she have expected me," Edith would ask herself, "since she never invited
me?"). . . . But already the second furtive glance darted at her had enabled
Edith to grasp everything that was concealed by the complicated language
of M. de Charlus . . . the obvious secrecy and mischievous intention that
it embodied rebounded upon a young Peruvian whom Mme. de Mortemart
intended, on the contrary, to invite. But being of a suspicious nature, seeing
all too plainly the mystery that was being made without realising that it
was not intended to mystify him, he at once conceived a violent hatred of
Mme. de Mortemart and determined to play all sorts of tricks on her.
    —Marcel Proust, *The Captive*, trans. C. K. Scott Moncrieff,
      in *Remembrance of Things Past*, vol. 2 (New York: Random
      House, 1927), pp. 567–68.

Chapter 5

ACCOMPANYING DE-MONSTRATIONS:
THE PUBLICATION OF A DE-MONSTRATION AT
THE HEART OF THE ACTION OF GROUPS OF ACTORS

THE INTERVIEWS I carried out with researchers in fuzzy logic and artificial intelligence, especially those who participated in the debates over Elkan's theorem, give us a better grasp of the way Elkan's de-monstration and its substitutes were produced and put forward. The interviews bring to light the actions undertaken and the positions adopted by the protagonists as they sought to accompany the proofs and counterproofs. They make new dimensions of the de-monstrative work apparent.

More particularly, they make it possible to reconstitute, at least in part, the history behind an essential step in the dynamics we are studying: the publication of Elkan's text. We have already seen how Davis and Hersh contested one model for knowledge production in mathematics, a model according to which the publication of proofs would constitute both the immediate consequence and the univocal sign of their intrinsic correctness.[1] Given the available data, it will be useful to compare the testimony of these authors with the modalities of the publication process in Elkan's case.

HOW ONE AND THE SAME DE-MONSTRATION CAN BE
REJECTED AND THEN ACCEPTED FOR PUBLICATION

In early 1994, I met the author of a contribution to comp.ai.fuzzy that stood out by virtue of its unusually elaborate character. During our interview at his workplace in the United States, this researcher, whom I shall call John Carpenter, reviewed the steps in the process leading to the publication of Elkan's proof and its sequels.

Several months before our meeting, bibliographical research into fuzzy logic had brought me into contact with some of John Carpenter's work. During a major conference on fuzzy logic that I attended, John Carpenter

---

[1] See Davis and Hersh, "Rhetoric and Mathematics."

had been represented on several occasions as one of the major researchers in fuzzy logic, along with L. A. Zadeh. His name also frequently appeared on lists of members of fuzzy logic conference organizing committees, as well as in announcements of tutorials designed to train engineers in fuzzy logic.

John Carpenter was generally depicted as one of the first researchers to work on the use of fuzzy logic in a particular area of artificial intelligence; he was also widely viewed as one of the specialists in the "foundations" of that logic. The length and breadth of his activities in the field led him to assume a privileged role as spokesperson for fuzzy logic.

With the support of a group of colleagues, his participation in comp .ai.fuzzy had acquired a certain visibility with respect to other messages. The nature of his text, characterized by a federative approach and stamped with authority in the face of Elkan's critics, corresponded well to the place in the front ranks that Carpenter had been able to acquire in other contexts.

This observation clarifies the picture of the way de-monstrative work was carried out on the electronic forum. Rather than coming across as interchangeable individuals, the contributors turned out to have been endowed with unequal resources from the outset. This situation helped put them in more or less favorable positions for coming out on top in the debates. Different types of interventions were possible depending on the resources available, and certain authors were likely to be read and cited more frequently than others.

The unequal distribution of resources seems to have played a decisive role in the way the debates evolved. In particular, let us recall that the contributors appeared to possess differing abilities to articulate heterogeneous registers in the face of Elkan's skill in this area. Compared to other researchers in fuzzy logic, John Carpenter displayed a pronounced aptitude for juggling logical and technological registers and also political ones (for example, when he denounced the antagonism toward fuzzy logic of people who organized conferences on artificial intelligence).

Carpenter had honed his competence as a *mediator* in the course of serving as a spokesperson and in the context of his various professional activities. He had a solid knowledge base in logic, as I was able to observe during a personal interview. At the time, he was working on technological systems in the artificial intelligence department of a research institute. He thus had skills related to the manipulation of logical symbols and to the development of technological devices. He was capable of carrying out translations between corresponding registers, of dialoguing with academics and engineers working in numerous fields, and, finally, of mediating among them.

The institute that employed him handled a large number of research contracts, both civilian and military, in a very broad range of fields. Its members in the artificial intelligence department developed theorem provers (software designed to prove theorems), for example, or, like Carpenter, robot software. They were highly skilled in logical manipulations and in the development of new formalisms. Several of them participated in research seminars or taught courses requiring the mastery of logical skills in the computer science department of a nearby university. The students they trained, especially at the doctoral level, constituted a ready pool for recruitment at the institute. Indeed, many doctoral theses in computer science consisted in the development of new logical systems, in the demonstration of certain "standard" properties that characterized these systems, and in the articulation of this work with software development, often under contract.

The links between the institute's artificial intelligence department and the university's computer science department were thus quite numerous. The institute's considerable investment in the field of logical computer science paralleled the research being undertaken at the university, and the reputation the latter had in that field.

John Carpenter thus worked in an environment that could be described as logic-oriented. To the extent that his colleagues (some of whom were his interlocutors) carried out research on a great variety of logical formalisms and not simply on fuzzy logic, Carpenter was thus not without resources to engage in dialogue with partisans of other logic-oriented approaches to artificial intelligence and to play the role of mediator in the debates over Elkan's theorem.

At the time appointed for our interview, Carpenter received me in his office after greeting me at the entrance to the building in which he worked. Access to this building was strictly controlled: the doors were carefully locked, and a guard was posted at the main door. However, it would be inappropriate to draw the poetic conclusion that the exercise of logic is so fragile that it can only be undertaken in a highly protected environment. The protections applied equally to other sections of the institute, and they had to do more generally with work being done in the building on civilian and military industrial contracts.

On the office shelves, in addition to books on various technical fields, there were a certain number of texts on logic; Carpenter's capacity as mediator thus appeared in the very content of his library. On a worktable in the office, a computer with an Internet connection allowed access to the comp.ai.fuzzy newsgroup (Carpenter was kind enough to show me how this worked). The nature and arrangement of the objects presented in his office emphasized once again the *physical* character of the work of logical production, calling on the researcher's hands and eyes. Carpenter had read the messages addressed to the electronic forum, had drafted his counter-de-monstration of Elkan's theorem, and had sent his text to comp.ai.fuzzy while seated in front of his computer screen. Sitting at this same workstation, he described to me the way he saw some of the stages in the process leading to the publication of Elkan's proof, and what happened next.

During the interview, Carpenter noted that he had exchanged electronic mail with Elkan during the debates on comp.ai.fuzzy. According to him, Elkan had made contact after Carpenter had posted a counter-de-monstration of the theorem on the forum. Elkan offered complementary explanations of his approach, in a logic of appeasement, and referred to the helpful encouragement he had received from major figures in artificial intelligence.

Carpenter remarked that Elkan must not have expected such reactions and needed to "correct his aim." During our interview, Carpenter offered some details about the content of their correspondence. According to Carpenter, Elkan declared that he had sent his article to AAAI'93 after it had been rejected by the program committee of a conference on fuzzy logic that we shall call FUZLOG. Carpenter asserted that Elkan had submitted his article to the FUZLOG conference, which Carpenter himself chaired at the time. Elkan's text had been rejected, and the author then sent a letter of protest to the chairman of the FUZLOG program committee (I shall call him Peter Smith), who helped John Carpenter organize the conference. Smith then replied to Elkan that he could write to Carpenter if he wanted further explanations about the rejection of his article. But, according to Carpenter, Elkan did not contact him at that point; he did so only after Carpenter's message appeared on the electronic forum.

Carpenter added that he had recently stopped corresponding with Elkan, after the latter had asked him for the names of the FUZLOG reviewers who had expressed negative opinions about his article, so he could get in touch with them. Carpenter justified his silence by explaining that he was concerned about preserving the reviewers' anonymity.

Some time after this exchange, I sent Elkan an e-mail asking him for some details on the matter. He explained that he had indeed submitted his article to the FUZLOG conference before submitting it to AAAI. His text was rejected by the FUZLOG program committee without comment. According to him, he then took the initiative of writing to the FUZLOG committee to try to get an explanation of the decision. He specified that as he saw it, an approach of that sort was perfectly legitimate. Moreover, in response to one of my questions, he explained that he had written his article completely independently, without having been drawn into the enterprise by "powerful" figures in artificial intelligence. Moreover, he argued that the very notion of "powerful" figures was debatable.

To justify the rejection of Elkan's article by the FUZLOG program committee, John Carpenter adopted an objectivizing register during our interview. For him, the very notion of logical equivalence utilized in Elkan's proof had led its author into error, for it could not be used without introducing a complete logical system. According to Carpenter, Elkan had confused two notions of truth: truth in the sense of truth *value*, and truth in the sense of a tautological statement based on a logical equivalence. In Carpenter's view, Elkan's approach was comparable to asserting that non-Euclidean geometry is impossible, while adding to the corresponding axiomatics an axiom of Euclidian geometry. In this view, then, Elkan had thus committed an elementary error; at best, he had made a trivial statement announcing a result that "everyone" had known for a long time, one that he could have laid out in two lines. Still according to Carpenter, Elkan had very little familiarity with the workings of thousands of applications of fuzzy logic. Carpenter summed up the situation by declaring that Elkan's article had been rejected for "obvious reasons."

Carpenter specified that his own efforts to write a counter-de-monstration did not mean that he thought Elkan's criticism was fruitful. He had deemed it important to respond to Elkan, he said, because the latter's article had won a prize during the AAAI conference and because the resulting promotion of his text threatened to tarnish the image and credibility of fuzzy logic.

In a parallel move, Carpenter had contacted the AAAI conference organizers to ask that works on fuzzy logic be presented and discussed during the conference. He denounced the partiality of the publication and promotion of Elkan's text when no other article on fuzzy logic had been retained for presentation at the conference.[2] He evoked the hostility of the AAAI organizers toward fuzzy logic; as proof, he mentioned the fact that none of the conference's themes corresponded to the category "fuzzy logic." For Carpenter, one source of the antagonism was easy to identify. The rise of fuzzy logic was an increasing threat to the financing of classical artificial intelligence. On the occasion of FUZLOG, nearly 500 articles reached the program committee, that is, as many as were submitted to the AAAI'93 conference—which, he noted, had nevertheless published no article on fuzzy logic but Elkan's.

To illustrate his claim, he recalled a sign that was, in his eyes, flagrant evidence of hostility toward fuzzy logic on the part of the organizers of major conferences on artificial intelligence. Carpenter drew on his own

---

[2] Except perhaps for a short presentation on the way a fuzzy controller works. See Alessandro Saffiotti, Nicholas Helft, Kurt Konolige, John Lowrance, Karen Myers, Daniela Musto, Enrique Ruspini, Leonard Wesley, "A Fuzzy Controller for Flakey, the Robot," *Proceedings of AAAI 1993* (Cambridge: MIT Press, 1993), p. 864.

experience in referring to an article he had submitted one year to a major conference on artificial intelligence (which I shall call AIC). His article was rejected. Carpenter believed that the decision had been made without a serious reading of his text. He explained that the first reviewer had justified the rejection by simply declaring that the article was not in the field of decision theory. To submit his text to AIC, Carpenter had been required in fact to situate it among a restricted list of thematic categories that did not include fuzzy logic. This choice played a determining role in the transmission of his article to a particular subcommittee. Carpenter chose the category that seemed to correspond best to his text, but he deemed the rejection of his article unfair because it had been preorchestrated by the absence of a category corresponding fully to his research.

The injustice seemed to him all the greater in that it had been relayed by a second justification for the rejection of his article, from a second reviewer—and this one was just as unwarranted, in Carpenter's eyes. This second response simply asserted that the technological device described in Carpenter's article "couldn't work." Carpenter noted ironically that this judgment, which he found quite peremptory, had been disavowed very quickly at a later meeting of the AIC. In fact, the project that he had presented in his article placed second at the robotic competition held during that later conference. The modalities of rejection of his article at an AI conference thus illustrated perfectly, for Carpenter, the hostility toward fuzzy logic manifested by the organizers.

Carpenter's testimony, along with Elkan's, made it possible to bring to light several dimensions of a work of *accompaniment* of the demonstrations. I am using the term "accompaniment" to characterize an approach that consists, for the actors, in repeatedly becoming spokespersons for their earlier de-monstrations, by reformulating their theses orally or in writing—that is, by producing new de-monstrations—and by involving themselves in various interactions that lead to advancing their own viewpoints. The accounts presented highlight the considerable activity—comparable to diplomatic efforts—in which the authors of proofs must engage in order to defend their work.

In order to support his de-monstration, Elkan did not limit his efforts to written interventions on the electronic forum. He had to adapt his demonstrative registers to those parties who contacted him or whom he took it upon himself to contact. In other words, Elkan did not economize his energy: he produced personalized responses after drafting a general response on the forum. Whether in order to create or to integrate new registers in his de-monstrative repertory, Elkan displayed as much flexibility in his personalized responses as he did on the electronic forum.

This flexibility, this differentiation of the registers used according to the interlocutors,[3] and the corresponding inflation of the meanings given to Elkan's undertaking in the course of his explanations, participated fully in the defense of his original de-monstration. They were part of an extension of his earlier de-monstrative actions. As surprising as it may seem, the reassuring message Elkan sent to a researcher in fuzzy logic as critical as John Carpenter and the declarations he made to me concerning the independence of his work must be counted in the long list of actions serving to advance his theorem.

Like Elkan, Carpenter also used personal contact and personalized explanations, or what might be called *adjustments*, to accompany his de-monstrations that were originally aimed at groups. He engaged in multiple interactions, entering into dialogue with Elkan, Peter Smith, and me. His involvement was also manifested by the letter he sent to the AAAI organizers.

Compared to Elkan, Carpenter used a more stable battery of arguments to support his counter-de-monstration. The criticisms he directed at Elkan's proof, like his denunciation of hostility to fuzzy logic on the part of the AAAI organizers, remained the same as the ones he expressed in a text addressed to the electronic forum. This consistency was found even in the use of a discourse on self-evidence, one that we have already observed in multiple interventions on comp.ai.fuzzy.

With Carpenter, as with Elkan, de-monstrative work was thus not limited to a dialectic involving the writing and reading of proofs, but also proceeded from involvement in a set of interactions in various modes. Supporting or advancing a de-monstration could be accomplished through undertakings as varied as sending a critical letter or a reassuring electronic message, asking for written justifications for the rejection of an article, or agreeing to engage in private dialogue about one's de-monstrative approach.

In fact, given the disagreements expressed about the more or less acceptable character of these undertakings, we see more clearly here that the actors' representations of the degree of legitimacy of the registers and actions utilized in the exercise of logic were quite variable, when they did not constitute an object of negotiation. In other words, it appears clearly that one dimension of the work of each participant consisted in defining a border (in the facts about which consensus was lacking) between what might or might not be associated with the activity of logic.

These exchanges thus inflect the way the process by which Elkan's theorem was made public can be represented. This process was not limited to

---

[3] See the citation from Proust at the beginning of part 3.

an unproblematic publication of a text, followed by debates in an electronic forum. From beginning to end, the dynamics involved proceeded from a whole series of heterogeneous actions. In the wake of the fruitless effort that preceded the rejection of his text by the FUZLOG program committee, Elkan undoubtedly benefited from the support of colleagues in the publication and the promotion of his text in the context of AAAI'93. But he must have been quite isolated as he faced his critics, at least in a first phase, and this presumably led him to multiply his efforts at "explanation."

The list of actions undertaken includes an apparent attempt to stanch the flow of criticisms formulated about his article in comp.ai.fuzzy by drafting a message addressed to the electronic forum. All else being equal, the confident tone Elkan adopted may have played a decisive role in the formation of a favorable opinion toward his de-monstration, in particular where "undecided" readers were concerned. In contrast, Elkan may have deemed an especially reassuring tone better suited to a message addressed to one of his most virulent critics.

This apparent mobility in Elkan's presentation of himself and his work, depending on the addressee and on the interactions in the context of which it was manifested, reveals in passing that de-monstrative work can be grasped in part as a specific moment in the staging of everyday life, to use Erving Goffman's terms.[4]

A Goffmanian approach seems equally well suited to account at least in part for the actors' negotiation of the conditions of their interactions. In fact, the actors in the Elkan controversy put forward contradictory standards for the conduct of a dialogue, in order to establish or to break down forms of conciliation. It was in this specific context that Carpenter broke off all dialogue with Elkan, by evoking the "rule" according to which the anonymity of reviewers must be respected.[5]

Still, it must be emphasized that the logical texts produced by the authors in question constituted essential entities around which interactions were constructed. These interactions were not established simply between individuals. They were also established between human beings and things: the protagonists interacted with texts, computer screens, and keyboards. As we have seen, logical texts, which can be compared to instruction manuals for technological devices, given the lists of instructions that they provide for their own reading, also played a role in the encoding and partial determination of the conditions under which readers interacted with these

---

[4] See Goffman, *Presentation of Self*, and Erving Goffman, *Forms of Talk* (Philadelphia: University of Pennsylvania Press, 1981).

[5] Let us note that Steven Shapin's research on the role of the establishment of and respect for behavioral codes in the institution of scientific credibility supplies another illustration of the value of using an approach of the Goffmanian type in the history and sociology of the sciences. See Shapin, *Social History of Truth*.

texts. Consequently, if a Goffmanian reading allows us to understand how the actors' negotiation of ways of doing logic was inscribed in a more general negotiation of ways of interacting, it is nevertheless appropriate to note that grasping the specificity of the production of logical texts allows us to understand the specificity of the interactions put into play, interactions that are not instituted solely among individuals.

Moreover, this description of the dynamics of elaboration and deployment of de-monstrations does not apply to Elkan's case alone. John Carpenter, too, carried out de-monstrative actions of various natures. And the stability of Carpenter's arguments constituted, as such, a specific form of presentation of self and of his de-monstration.

In addition, the undertakings of the other actors, particularly with regard to the publication of Elkan's text, must also be taken into account. The actual publication of the article in question appears to have resulted from a series of actions carried out by a multiplicity of participants, whether on the occasion of the possible "encouragements" brought to the development of the text, its acceptance for publication, or its promotion via the prize it was awarded. It is plausible that Elkan benefited from the support of a specific collectivity (one including scientific "personalities" among others, even if these latter were not "powerful"), support that contributed to the legitimacy and visibility of his de-monstration. While he failed to obtain the support of FUZLOG's representatives and evaluators, he did find support in the context of AAAI'93. He thus passed through the trials that determined the fate of his proof, one step at a time.

The episode we have identified as the publication of Elkan's text was thus quite complex. Its unfolding was linked in particular to the announcement of significant stakes (the future of the financing of research in artificial intelligence and in fuzzy logic) and to a conflict between scientists belonging to what Davis and Hersh would call two "subdisciplines," if not two "corporations."[6]

The accounts by John Carpenter and Charles Elkan thus allow us to pinpoint some of the mediations that came into play in the publication of Elkan's de-monstration, mediations that can be usefully compared to Davis and Hersh's descriptions. Davis and Hersh asserted, as we have seen, that a mathematical text submitted for publication could be the object of either a favorable or an unfavorable review depending on the

---

[6] During the interviews I carried out in the United States, a number of accounts offered by researchers in artificial intelligence converged to evoke the considerable decrease, during the 1980s, of military financing for "traditional" artificial intelligence in favor of neural methods. The fear on the part of researchers in "traditional" artificial intelligence described by John Carpenter—fear that they would lose more and more sources of funding with the rise of "new methods" such as fuzzy logic or neural networks—was thus perceptible in the discourse of the "opposing camp," and was associated with precedents.

journal to which it was submitted. The case of Elkan's proof provides an excellent illustration of their thesis.

Elkan's de-monstration was considered correct by some and incorrect by others. Various factors seem to have influenced the decisions made in one place or another to publish his de-monstration or not. These decisions were not based simply on a verification of respect for universally shared norms throughout the steps in Elkan's proof, norms that would have made it possible to reach consensus as to the proof's correctness "automatically." The decision to publish Elkan's result was not a de facto "guarantee" of its "correctness," nor did the decision not to publish confirm its "absence of correctness" in some ideal sense. While sets of criteria deemed universal were used by the AAAI and FUZLOG reviewers to assess the proof's correctness,[7] these criteria clearly did not suffice to produce consensus.

Here we see a "subdiscipline" effect in the sense in which Davis and Hersh use the term. These authors suggest that one has to look beyond the image of neutrality of formalisms[8] in order to identify the tacit skills brought to bear locally, skills that simultaneously affect reviewers' decisions about the correctness of a proof, make these same decisions opaque for a "nonspecialist," and help determine which considerations are deemed relevant (for example, what "poses a problem" or what is "interesting," to use Davis and Hersh's terms).

The data available up to now have not allowed me to test the thesis put forward by Davis and Hersh according to which reviewers generally carry out only partial verifications of proofs and take the remainder of the demonstration on trust, based on the author's reputation, that is, on the reliability and originality he or she has demonstrated previously. In contrast, we have had a confirmation of the role played by the reviewers' representation of the research field covered by the forum that solicited them, in the decision to accept or reject an article.

It may be surprising for some readers to learn that articles in the fields of mathematics, logic, and artificial intelligence are subjected to such practices of evaluation. In fact, the testimonies of John Carpenter and Charles Elkan allow us to draw a somewhat different picture of the reviewers' work than the one Davis and Hersh provide. This work turns out to be very similar to the work of the authors of the articles themselves. Like the authors, reviewers produce written texts; like the authors, reviewers must de-monstrate the well-foundedness of their own points of view. And then,

---

[7] Criteria presented as universal, although in fact not shared, were evoked to justify the decisions made.

[8] But also beyond the idea that some universally shared criteria would make it possible to end up with a consensus about the correctness of the proofs.

as we have seen, challenging an evaluation may lead to an inflationist production of new de-monstrative registers.

This analysis makes available a certain number of elements that enable us to begin to grasp the dynamics that led to the publication of Elkan's de-monstration. But the accounts offered by Elkan and Carpenter raise numerous questions. The first of these has to do with the modalities of the process whereby Elkan developed his de-monstration. Is it the case that some actors played a role in the phase that preceded the submission of Elkan's article to the conferences in question? And, if this is the case, can this role be circumscribed more precisely? To answer such questions, we shall need to analyze additional testimony.

From De-monstration to Publication: The Importance of Interactions

As noted earlier, I met Charles Elkan on the University of California, San Diego campus when I was attempting to assess the diversity of forms of logical activity that could be found within that university. During my inquiry, a specialist in cognitive science advised me to get in touch with Elkan, an assistant professor in computer science who was doing research in logic. I made an appointment via e-mail to meet him in his office one day in early 1992. Before we analyze the account he provided on this occasion of the process by which his de-monstration had been developed, we need to review the nature of his activities at UCSD in some detail, as this will make it easier to appreciate the modalities of his de-monstrative work.

Like many of his colleagues with whom I had some contact, Elkan did most of his research from a *workstation* in an office in the university. Like John Carpenter's, this workstation consisted principally of a table furnished with a computer connected to the Internet. In this space, Elkan could *draft* his articles, *connect* to the electronic forum comp.ai.fuzzy, and *read* various texts on screen or on paper. In this framework, he could also *interact* with colleagues, students, or visitors such as me; however, his space for interactions was by no means limited to this particular site. I was able to observe over time that his space for interactions extended to classrooms, seminar rooms, and even cafés. Indeed, following our first meeting, I had occasion to see Elkan a number of times during seminars that he had organized or in which he was participating, and also to talk with him in private or in the presence of third parties in his office or in cafés. I want to stress the fact that Elkan's interpersonal activity took place as a complement to the solitary work that was carried out in his office.

The seminars Elkan organized or in which he participated were varied. They had to do in large part with the development of computer methods or projects in artificial intelligence, chiefly under the auspices of the UCSD computer science department or the cognitive science department, this latter being very active in the field of research on neural networks. Other seminars involved members of other departments. For example, Elkan organized a seminar on Heidegger in collaboration with a specialist in analytic philosophy, Adrian Cussins,[9] and a researcher in cognitive science and communication based in the Department of Communication, Philip Agre.[10]

Elkan attended various talks organized by the UCSD science studies program, which brought together philosophers, historians, and sociologists of science. At the time, the program was directed by Gerald Doppelt, Robert Marc Friedman, and Chandra Mukerji; its members included Philip Kitcher, Bruno Latour, Michael Lynch, and Steven Shapin.[11] Among others, Elkan had attended a lecture by David Bloor on the sociology of "2 + 2 = 4," as well as a lecture by Joan Fujimura[12] on projects for decoding the human genome (Elkan himself was interested in the development of computerized methods for dealing with data generated by this type of project). After Fujimura's talk, Elkan explained to me that he was interested in the sociology of science in relation to his own work (this is what led him to support my research). According to him, scientists themselves could and even should contribute to the development of the sociology of science, although he personally found it more interesting to "do science" than to observe it.

Thus Elkan developed a capacity for dialogue with a wide range of interlocutors. Viewed in retrospect, this ability corresponds well to the multiplicity of registers utilized in his article and on the electronic forum. Elkan's training as a researcher in computer science had probably played a major role in honing his competence as a *mediator*, that is, as someone with an aptitude for making connections among results in a large number

[9] See Adrian Cussins, "Content, Embodiment and Objectivity: The Theory of Cognitive Trails," *Mind* 101, no. 404 (1992): 651–88.

[10] Philip Agre and David Chapman, "Pengi: An Implementation of a Theory of Activity," in *Proceedings of AAAI-87: The Sixth National Conference on Artificial Intelligence*, ed. Kenneth D. Forbus, and Howard E. Shrobe (Los Altos, Calif.: Morgan Kaufman, 1987), pp. 268–72.

[11] See Gerald Doppelt, "Kuhn's Epistemological Relativism: An Interpretation and Defense," *Inquiry* 21 (1978): 33–86; Robert Marc Friedman, *Appropriating the Weather* (Ithaca, N.Y.: Cornell University Press, 1989); Chandra Mukerji, *A Fragile Power: Scientists and the State* (Princeton: Princeton University Press, 1989); Philip Kitcher, *The Nature of Mathematical Knowledge* (New York: Oxford University Press, 1984).

[12] See Adele E. Clark and Joan H. Fujimura, eds., *The Right Tools for the Job: At Work in Twentieth-Century Life Sciences* (Princeton: Princeton University Press, 1992).

of fields, given the diversity of subjects taught within the framework of undergraduate majors and graduate-level masters and doctoral degree programs in computer science.[13] His international experience may also have had something to do with the acquisition of such capabilities. Before being hired as an assistant professor at UCSD, Elkan had earned his bachelor's degree at Cambridge University in England, had written his Ph.D. thesis in computer science at Cornell University, and had had a postdoctorate appointment at the University of Toronto.

Elkan's research activity could be seen as an extension of a quite specific professional commitment. While he distinguished among several trends in artificial intelligence, in particular in relation to their attachment to a pro- or antilogic approach, Elkan indicated that he was personally interested in a logic-oriented approach to "traditional" artificial intelligence (thus invoking a category used by Carpenter and other contributors in messages addressed to the electronic forum).

In particular, Elkan was attempting to draw up, on paper and by computer, some general procedures for solving dynamic decision-making problems—"canonical" problems in artificial intelligence.[14] The "solutions" Elkan developed relied on so-called first-order classical logic, or a logic known as nonmonotonic.[15] These choices brought him into close proximity with researchers working in northern California, at Berkeley and Stanford, where he was interested in pursuing his own research.[16]

In keeping with John Carpenter's claims, my own observations revealed that researchers with similar interests manifested various forms of hostil-

[13] This phenomenon is not at all surprising, given that computer science is itself involved in a very large number of scientific activities.

[14] Such as the "Qualification Problem" or the "Frame Problem." See Charles Elkan, "Reasoning about Action in First-Order Logic," in *Proceedings of the Ninth Biennial Conference of the Canadian Society for Computational Studies of Intelligence* (CSCSI'92), pp. 221–27 (Vancouver: Morgan Kaufman, 1992); Elkan, "The Qualification Problem, the Frame Problem, and Nonmonotonic Logic," working paper, University of California, San Diego, 1992.

[15] See the two articles cited in note 14, and also Charles Elkan, "Formalizing Causation in First-Order Logic: Lessons from an Example," *Working Notes of the AAAI Spring Symposium on Logical Formalizations of Commonsense Reasoning* (Stanford, Calif.: Stanford University, March 1991), pp. 41–47; Elkan, "Incremental, Approximate Planning," in *Proceedings of the Eighth National Conference on Artificial Intelligence (AAAI-90)* (Boston: AAAI Press, MIT Press, 1990), pp. 145–50; Elkan, "Conspiracy Numbers and Caching for Searching and/or Trees and Theorem Proving," *Proceedings of the Eleventh International Joint Conference on Artificial Intelligence* (Aug. 1989), pp. 341–46; Charles Elkan and David McAllester, "Automated Inductive Reasoning about Logic Programs," in *Fifth International Conference and Symposium on Logic Programming* 2, ed. Robert A. Kowalski and Kenneth A. Bowen (Aug. 1988), pp. 876–92.

[16] Elkan gave me a photocopy of the dossier he had established as a candidate for a position at the University of California, Berkeley: Charles Elkan, "Research on the Qualifi-

ity to fuzzy logic.[17] Thus, Elkan's research and his critical attitude toward fuzzy logic were by no means isolated phenomena; on the contrary, they were solidly connected to the set of choices, practices, and relational networks within which he was operating. Nor was the fact that Elkan carried out research in logic within a computer science department an isolated phenomenon. According to Elkan, very few people with doctorates in logic could continue to work in that field inside a department of mathematics or philosophy. As he saw it, finding a position in an American university was much more difficult for someone with a Ph.D. in mathematical logic than for people with doctorates in any other scientific discipline. He reported that Ph.D.s in logic generally had no way to earn their living except to work, as he was doing, in the field of computer science, or else to teach in a high school.[18]

In our initial conversation, as I was asking about his research, Elkan told me that he was in the process of writing an article on fuzzy logic and would send me a draft copy. This text, a first version of the article published in the proceedings of AAAI'93, presented a different thesis, one denouncing the inconsistency of fuzzy logic. When I asked him to explain what had led him to write the article, Elkan said that he had become acquainted with fuzzy logic by reading several newspaper and magazine articles that set forth its basic principles and discussed its "applications" in the realm of household appliances. "Intrigued" by what he had read, he had consulted an article in an academic journal written by an author presented as one of the principal specialists in fuzzy logic, Bart Kosko. Then he had participated in an international conference devoted to fuzzy logic.

---

cation and Frame Problems," Miller Research Professorship Application, University of California, San Diego, January 1992.

[17] As an illustration, we may consider an account offered by one of the major figures in "logic-oriented" artificial intelligence in the United States. During a seminar I attended in 1994, this renowned scientist made the following statement in response to a question: "Now with regards to fuzzy logic, I have to admit a reaction to it which is similar to the composer Rossini's reaction to Wagner's operas. Somebody asked him what it was, and he said: 'This is a work one can't make a decision about for the first year, and I have no intention of listening to it for more than a year.' So I worked a little bit in fuzzy logic and this discouraged me for looking at it anymore. . . . There was a contest between Hitachi and Mitsubishi about the supply of elevators. And Hitachi's elevator control system used fuzzy logic. And its rival's was just an ordinary control system, and the system using fuzzy logic was much slower. It maintained my opinion." These remarks provide further evidence that diverse demonstrative registers can be used to characterize and condemn a logical system. Let us note in particular how technological devices can be invoked to criticize a given formalism.

[18] Having been unable to unearth any sociological studies on the profession(s) of logician—to the extent that such a notion has any meaning—I am in a position neither to support such an assertion nor to refute it. Still, my observations do not contradict such a statement; they tend rather to reinforce it.

The first piece he had read on the subject had appeared in *The Economist*, in a section devoted to science news. The article presented the contradictory viewpoints of two researchers on fuzzy logic. The first was enthusiastic, and referred to Kosko's essential work on the subject. The second, on the contrary, was very critical. Elkan noted that he "already believed" more in the relevance of the second analysis than in that of the first.

An article by Kosko had been cited in the context of that "discussion," and Elkan located a copy at the UCSD library.[19] In this text, he believed he had identified the axiomatics that he attributed to fuzzy logic and that constituted the point of departure of his own article—although the specific system of axioms in question was not explicitly evoked by Kosko. Elkan said he was aware that there were hundreds of books and articles on fuzzy logic, but he acknowledged that he had not read them and that his representation of fuzzy logic was based primarily on Kosko's article.

When I asked him if his own hypotheses on the axiomatic system of fuzzy logic might not be challenged by authors who had proposed other definitions for fuzzy logic, he replied that such a situation was always conceivable. According to him, the option of defending fuzzy logic by proposing new definitions for the latter constituted a "game" that could be played ad infinitum. He declared, however, that he had learned nothing further about the "theory" of fuzzy logic during the international conference in which he had participated. He recalled one particular talk prepared jointly by two specialists in fuzzy logic; he had judged it "completely incomprehensible." He also referred to the invitation proffered by some of his UCSD colleagues to Bart Kosko, whose recent talk Elkan had attended.

Having assessed the definitions and de-monstrations of the properties of fuzzy logic that he had thus identified, Elkan criticized that logical "theory" as such, judging that its "applications" worked well but that these successes should not be attributed to fuzzy logic per se. He thought that the "fundamental idea" developed in fuzzy logic was the use of the operators "min" and "max" to make decisions in situations of uncertainty. According to him, this approach was fundamentally flawed, because it did not distinguish between dependent and independent propositions; in his view, these could be handled only by probability theory. As Elkan saw it, the axioms of fuzzy logic that used the operators min and max represented hypotheses that were "too strong" in comparison with those used by probability axioms; this distinction accounted for the inconsistency of fuzzy logic.

[19] Kosko, "Fuzziness vs. Probability."

According to Elkan, if, as an experiment, one were to bring together a probabilist and a fuzzy logician in a game of chance and have them engage in a series of bets (each player would develop a strategy corresponding to the mode of reasoning he defended), the probabilist would end up winning and the fuzzy logician would lose. Elkan asserted that this experimental criterion was a decisive argument. In his view, the probabilists had demonstrated once and for all that the only way to reason about chance was to use probability theory. Elkan affirmed that the applications of fuzzy logic had a certain practical value, but that the applications that "worked" did not "really" use fuzzy logic.

I then asked Elkan if he had submitted his analyses to researchers in fuzzy logic. He replied that he had simply asked for some "clarifications," deeming it useless to go further. In contrast, he said that he had benefited from the opinion of an important figure in probability theory, who had been the sole reader of the proof of his own theorem up to that point.

According to Elkan, this probabilist declared that he had "suspected" for a long time that such a result could be demonstrated, and that Elkan's proof had confirmed his own doubts. This comment had in turn reinforced Elkan's conviction that his demonstration was "flawless." He added that while his interlocutor had recently published a collective volume about reasoning under uncertainty that included an article on fuzzy logic, it was only because the publisher had asked him to do so. As Elkan recalled, the probabilist had told him that the article on fuzzy logic he had chosen was the one that he "understood best," even though he "did not understand everything." Elkan made it clear that this formulation was ironic, that it signified in a veiled way that the article was "incoherent," like fuzzy logic itself.

According to Elkan, probabilists generally formulated criticisms of fuzzy logic identical to his own, and in particular the following: it has been demonstrated once and for all that the only way of reasoning about chance stems from probability theory. But he said that these criticisms were mostly formulated orally, rarely in writing; when a few probabilists got together, it was customary to make jokes about fuzzy logic.

Moreover, Elkan claimed that critical attitudes toward fuzzy logic were quite commonplace in the scientific community. As evidence, he noted that only one major American university had a researcher working "officially" on the *theory* of fuzzy logic, someone I shall call Michael Jones. According to Elkan, Jones already held the rank of professor when he had begun to work on fuzzy logic; this move had brought him only loss of respect from his colleagues. Elkan thought that Jones had failed to convince anyone in his own university, with the result that the position opened up by his retirement was to be given to a researcher working on reasoning under uncertainty, to the *exclusion* of fuzzy logic. While certain researchers working on neural networks believed that fuzzy logic was valuable and were developing neurofuzzy methods, Elkan had very critical things to say about those methods. According to him, some specialists in neural methods even believed that it was "senseless" to combine the two approaches.

I then questioned Elkan about the motives that had led him to draft his text. He replied that he hoped his article would at least allow some individuals to become aware of the limits of fuzzy logic so that they could take their research in more fruitful directions. He was eager to complete his draft quickly, but he felt that something was still missing. As a complement to his critiques, he wanted to evoke more "positive" aspects and to explain that, while the theory was inconsistent, certain of its "applications" worked "well." He compared fuzzy logic to N-rays, judging that fuzzy logic would doubtless meet the same fate and fade away, perhaps more slowly but just as irrevocably.[20]

In the light of this account, which constitutes a personalized de-monstration, or one *adapted* to the interlocutor (me, in this instance), the dynamics of elaboration of Elkan's theorem loses a little of its opacity, and the image of the work of proof emerges transformed at the same time. The preparation phase of Elkan's article, which might have been viewed a priori as a solitary process inaccessible to sociological investigation, appears on the contrary to have been marked by a set of decisive interactions. As his article progressed, Elkan's growing confidence in the correctness of his proof (or rather proofs—we shall come back to this point) was nourished by his observations of other convergent viewpoints. As we have seen, De Millo, Lipton, and Perlis stressed the determining role, in the construction of proofs, of the interactions in which the authors were involved.[21] In Elkan's case, these interactions are of several orders.

First of all, we can observe the effects of Elkan's interactions with specific texts. His reading of an article that developed critiques of fuzzy logic clearly confirmed in his own mind the correctness of his own judgments of this logic, and thus helped strengthen the positions he adopted. The interventions of various actors also appear to have taken on considerable importance. The success Elkan encountered when he *tested* the rough draft of his proof obviously constituted a supplementary guarantee reinforcing his conviction that his proof was "logical." The fact that he

---

[20] In the spring of 1903, the French physicist René Blondlot reported that he had discovered a form of radiation that he called "N-rays." In the fall of 1904, the American physicist Robert William Wood, after visiting Blondlot's laboratory, published a report in *Nature* that was largely viewed as a proof that N-rays did not exist. For a historical account that casts doubt on the well-foundedness of Wood's demonstration contesting the existence of N-rays, see Ashmore, "Theatre of the Blind."

[21] De Millo, Lipton, and Perlis, "Social Processes."

viewed this early reader as an eminent probabilist must only have increased the value of the latter's approval.

Thus Davis and Hersh's testimony on the importance of the lack of counterevidence in the dynamics of sedimentation of proofs finds a specific and precise empirical confirmation here.[22] We can now better identify the nature of the "encouragements" on the part of "personalities" from which Elkan said he benefited.

Several other researchers undoubtedly played roles similar to that of the probabilist throughout the process that led to the publication of Elkan's text. At all events, the acknowledgments Elkan included at the end of his article support this hypothesis: "The author is grateful to several colleagues for useful comments on earlier versions of this paper, and to John Lamping for asking if Theorem 1 holds when equivalence is understood intuitionistically."[23]

The effect of numbers in the dynamics of mutual strengthening of viewpoints is perceptible here. We note once again that the observation of convergences helps transform into *conviction* what had once belonged, for the actors, to the realm of the *uncertain* or the *intuitive*. In the exchanges on the electronic forum, as we have seen, one participant explained that he had "had the same thought" as one of Elkan's critics, without being as sure of it *then* as he was after reading the latter's message. In other words, we see how interactions can help modify the status attributed *collectively* to proofs, and how their correctness can be experienced in terms of degree of conviction, influenced by the test of numbers.[24]

We find another illustration of this phenomenon, moreover, in Elkan's reprise of oral criticisms of fuzzy logic. Elkan takes into account "informal" declarations and convergent attitudes in formulating his own viewpoint. The habitual jokes of certain probabilists, the charges of absurdity leveled by specialists in neural networks, and even the critical attitude toward fuzzy logic of Michael Jones's colleagues are all perceived and brought forward as possible evidence, from Elkan's standpoint and that of third parties, for the appropriateness of Elkan's undertaking.[25]

[22] See Davis and Hersh, "Rhetoric and Mathematics."

[23] Elkan, "Paradoxical Success," August 1993, p. 702.

[24] The expression "test of numbers" does not correspond to the same reality as the one described by the expression "argument from authority," from my standpoint, which is an extension of Bruno Latour's associationist analysis in *Science in Action*, p. 31.

[25] In 1994, I interviewed a probabilist from the university where Michael Jones worked; this researcher expressed criticisms identical to the ones Elkan attributed to certain probabilists. He judged in particular that the operators min and max were not suitable for reasoning under uncertainty. A probabilist from another major American university formulated similar criticisms. I had questioned him on this subject in 1992. A few years earlier, he had invited Jones to present fuzzy logic in his seminar; several sessions had been devoted to Jones's presentation of fuzzy logic.

The same thing holds true for what Elkan perceives and displays as concordant *institutional signs*. Such is the case for the nonattribution of jobs to theoreticians of fuzzy logic in the most important American universities, or for the anticipated awarding of Jones's vacant chair to a researcher who does not work on fuzzy logic.

These elements carry all the more weight inasmuch as Elkan's dialogue with researchers in fuzzy logic is quite limited and does not dwell on the diversity of the approaches developed by the latter in an abundant literature. This posture on Elkan's part itself helps reinforce his conviction of the appropriateness of his own representations of fuzzy logic. It has its context in a practice shared by a large group of actors. The principal de-monstrative register that underlies it among his colleagues ultimately proceeds for the most part from an argument from authority. When a recognized researcher asserts that he or she is unable to understand work carried out in fuzzy logic, the interlocutor is presented in effect with the following alternative: the latter must conclude either that the researcher's reputation is unwarranted or that the theses advanced by the proponents of fuzzy logic are incoherent.[26]

Elkan's testimony shows that having to express himself in differing contexts leads him to formulate critiques that are evolving but also *differentiated*. In relation to his interlocutors and to the forums in which he develops his theses, Elkan uses distinct de-monstrative registers. As the formulation of his critiques and their de-monstrations are adapted to the individuals and groups he is addressing in turn, they take on an increasingly polysemic character.

To characterize fuzzy logic, Elkan used less nuanced qualifiers in our interview than those he used in his article or in his message to the elec-

---

[26] See once again the testimony cited in note 17, above, by one of the great figures in "logic-oriented" artificial intelligence. This researcher claimed that he no longer sought to understand the principles of fuzzy logic after having made some efforts to that end. Moreover, a probabilist working in a major American university made the following statement in an interview I conducted in 1992: "I have never heard anything at all clear on fuzzy logic. Michael Jones participated in my seminar, but I simply don't know what it was about. No statistician could tell you. It has existed for a long time, like parapsychology. I know that one has to replace {0, 1} by [0, 1], I know that some people use it, but I have never known why. But there are a lot of things I don't know. ... I spent quite a bit of time trying to understand; Michael Jones came to my seminar; I read introductions to fuzzy logic. But Michael Jones's colleagues, in his own university, don't take him seriously. He asserts that fuzzy logic constitutes a revolution, but I don't get it. It's a field that is evolving on its own. There may be something political to it, I don't know." By claiming here that he could not make any more efforts beyond those he had already made to "understand" fuzzy logic, this academic was invoking an argument from authority, since his interlocutor had then to opt for one of two interpretations: either this scientist had very limited capacity for judgment, or else fuzzy logic was itself incoherent. Moreover, my interlocutor himself brought up the prospect of a widespread condemnation of fuzzy logic in support of his own point of view.

tronic forum. At this stage and in this context, fuzzy logic was presented as an inconsistent theory. Furthermore, what Elkan described as constituting the "deep" reasons for the inconsistency of fuzzy logic[27] never came up again in his written interventions. Certain de-monstrative registers, seemingly customary and easy to call up in oral discourse, did not cross the threshold of writing or rereadings. Others, in contrast, appeared inescapable in the writing phase (Elkan alluded to this phenomenon when he asserted that he wanted to evoke more "positive" aspects as a complement to his criticisms).

Thus we can once again observe the flexibility Elkan manifested in his de-monstrative work. This aptitude was of course not unrelated to his pronounced abilities as a mediator. These capacities were manifested, for example, when he proved capable of using a highly targeted de-monstrative register in our conversation while referring to a "case" in the history of science (the "refutation" of the N-ray theory).

However, it is important to note that the production and use of differentiated de-monstrative registers do not stem from an isolated approach that would be peculiar to Elkan. In fact, they constitute a practice shared within a network of actors. We have already seen this in the way eminent scientists, in their interactions with *complicit* or *naive* interlocutors, used anecdotes, jokes, or other ready-made witticisms in order to *shine* in public, and in the way they displayed their own incomprehension or "ignorance" of fuzzy logic in order to discredit it, given the improbability that their own astuteness or erudition would be called into question.[28]

Elkan's testimony itself clearly showed an extensive recourse to targeted ranges of de-monstrative registers in relation to the nature of the interactions in which the actors engaged. Elkan thus stressed the custom, among probabilists, of evoking fuzzy logic more or less derisively in casual discussions in the corridors, and the habit of adopting a different discourse in writing (even agreeing to publish an article written by researchers in fuzzy logic if called upon to do so for editorial purposes).

The analysis of this testimony ultimately helps reduce the opacity of the "black box" representing the phase during which Elkan's proof was developed. Above all, it brings to light the essential role of the interactions in which Elkan was involved throughout the entire process. It also allows us a better grasp of the modalities of the de-monstrative work to which the actors devoted themselves, by stressing in particular a generalized recourse to differentiated de-monstrations.

---

[27] For example, the inappropriate character of the operators min and max for reasoning under uncertainty, or the idea presented as decisive according to which a probabilist bettor will always win against a "fuzzy" bettor.

[28] In particular, see the testimonies cited above in notes 17 and 26.

Still, many obscurities remain in the dynamics that later led to the rejection of Elkan's article for presentation at the FUZLOG conference and to its acceptance in the context of AAAI'93. It is time to look more closely at the way that essential phase unfolded.

ELKAN'S PROOF AND THE CONFERENCE
PAPER SELECTION PROCESS

While I could not observe the work of the organizers and reviewers of FUZLOG and AAAI'93, I was able to attend another large conference on fuzzy logic—I shall call it FUZCOL—and collect testimony on the way articles were selected for this event. The process is worth examining for several reasons. First of all, FUZCOL and FUZLOG, a year apart, brought together comparable groups of participants; their programs and protocols were very similar. Moreover, there is no reason to think that the practices of the two program committees differed fundamentally from one year to the next (some interviewees made explicit comparisons between the evaluation procedures used in various conferences, confirming the assumption that these did not vary significantly). It is thus likely that the general framework in which the FUZLOG rejection of Elkan's article came about can be grasped by studying the FUZCOL selection process. For similar reasons, it seems fair to suppose that this extrapolation can be applied as well to the general conditions under which the article was accepted for AAAI'93.

However, one cannot begin to grasp the nature of this process without first addressing a series of related questions. First of all, what led Elkan to submit his text to the program committees of these colloquia? What does such an undertaking signify, and what is at stake? What role does the selection of articles play in the unfolding of such colloquia? More fundamentally, in what do such gatherings of actors consist?

As I did not find satisfactory answers to these questions in the literature (the sociology of science included), I cannot avoid describing and analyzing the way FUZCOL worked in some detail, first summarizing the conditions under which the conference was organized and took place and then focusing on the way its organizers selected papers for presentation.

Let me point out right away that I had learned that FUZCOL was being held through a newsletter published by Écrin, a French association to which I belonged. Financed by the National Center for Scientific Research (CNRS), the Atomic Energy Commission (CEA), and some major French industrial groups, Écrin's goal was to "encourage contacts between research and industry." It was structured around thematic action groups, such as the "Crin environment club," the "Crin combustion club," and the "Crin fuzzy logic club." This last "club," created in 1991, brought together several researchers in fuzzy logic. It published a newsletter called the *Fuzzy Logic Club Letter.*

One issue listed several forthcoming conferences, and in particular that of FUZCOL in the United States. The notice specified that articles satisfying certain formal constraints could be submitted to its program committee. Paper proposals could be sent to the two presidents, whom I shall call David Belmont and Thomas Berger. Registrations, in contrast, were handled by a large association of American engineers, which I shall call USENG. By reading assorted brochures and analyzing various accounts, I was able to observe after the fact that several associations seeking to promote fuzzy logic throughout the world had made major contributions to the resources (posters and newsletters in particular) required for the organization of such a *gathering.*[29]

A few months before the conference, I contacted the USENG office in charge of registration. It was possible to attend the conference without giving a paper. I had only to send for a form, fill it out, and pay the registration fee. A certain amount of information was then mailed to me, mainly concerning means of access to the place where the conference was to be held.

I arrived on the first day of the conference, and it was immediately apparent that everything had been set up to support de-monstrations and interactions. FUZCOL took place in a vast hotel complex that offered numerous spaces where such practices could flourish.

The de-monstrative activity was carried out primarily in the context of the participants' communications. The term "de-monstration" is particularly fitting here, because the presentations of articles were located at the crossroads between the implementation of technologies of proof and recourse to various forms of ostentation. They put into play a linked series of moments devoted above all to *showing.* Thus one could find some of the same modalities for presenting symbolic proofs that were used in the electronic forum comp.ai.fuzzy.

The communications were the object of a very elaborate spatiotemporal organization. They were spread over four days, running from 8:00 a.m. to 5:00 p.m. As a general rule, apart from several plenary sessions,

[29] A registration form indicated that the conference had been organized with the support of several professional groups within USENG, plus several American, Japanese, and international groups supporting the development of fuzzy logic. A brochure prepared by the conference president listed talks to be given at the conference and mentioned that the neural networks group from USENG had sponsored this event. This apparently isolated commitment in fact echoed the many neurofuzzy projects presented at the conference. It also corresponded to the absence of a professional group specifically devoted to the theme of fuzzy logic within USENG.

three or four talks were given simultaneously. They took place in rooms equipped with microphones, overhead projectors, and projection screens. These venues were close together, which made it easy for listeners to move from one to another. The audience consisted for the most part of other speakers at the conferences.

The presentations were strictly limited in duration. Aside from the plenary sessions, which lasted about an hour, every paper was allotted twenty minutes. The first ten minutes were used for a short exposé, and the time remaining was devoted to discussion with the audience. The communications were grouped in thematic clusters, each involving four participants, for a total of ninety minutes. They were generally in English—or something close to it—with the help of an overhead projector. Introductory classes in fuzzy logic were offered for an extra fee the day before the conference began. John Carpenter, who was a member of the program committee, was one of the speakers in charge of these seminars.

In addition to the spaces set up for these specific forms of demonstrations, there was also a room set aside for "demos." The term "demo," used by the actors themselves, referred to software presentations, videos showing machines at work, or exhibits of various materials, such as electronic components, devices at rest or in action: for instance, a hammer moving along a metallic bar from which it was suspended. This mechanism was said to be driven by a *heuristic controller* that worked owing to fuzzy logic.

The object of the demos carried out by authentic business *representatives* was to *illustrate* the properties of fuzzy logic while promoting the sales of "fuzzy" products at the same time.

Several firms that had developed "fuzzy" products had set up *stands* where their representatives carried out a good deal of de-monstrative activity.[30] These spaces offered documentation on the products presented. The salespeople-demonstrators made general presentations showing how the devices worked, but they also entered into dialogue with participants, who were sometimes invited to try out the software themselves. During these demos, the *de-monstrators* attributed the effectiveness of the technological systems to the properties of fuzzy logic, turning themselves into commercial spokespersons or representatives, since commercial transactions were at stake. These de-monstrators spoke for the mechanisms by exhibiting—and elevating, as it were—the general properties of fuzzy

[30] Two companies presented software that worked "thanks" to fuzzy logic. Another exhibited both software and electronic components. It invited "curious" onlookers to carry out computer manipulations themselves. Two Japanese companies presented the virtues of electronic components by means of demos showing mechanical devices. Finally, one other company exhibited software and electronic components. It presented both demos carried out live on the spot and videos of demos.

logic. Presented as exportable to other systems, these properties became detached from the materiality and specificity of the mechanisms. In this movement of elevation, fuzzy logic and the products presented were thus promoted simultaneously.

Here again, the term "de-monstration" is well suited to characterize the practices observed. The activity put into play by these practices was partly *monstrative*, in the sense in which the word is being used in this book. But the qualifier does more than simply highlight the features common to various forms of exhibition. The term "monstrative' appears all the more justified inasmuch as the expression "public experiment" may be misleading to some readers as a synonym for demos. In an extension of work done in the history of science on experimental practices,[31] one might be inclined to use the latter formula spontaneously. However, to do so might confuse readers unfamiliar with the history of science as to the nature of an exercise that is closer to a de-monstration (a presentation prepared in its smallest details for specific occasions—hence the use by the actors of the term "demo") than to some of the everyday gropings characteristic of laboratory work.[32]

These demos were not designed to allow the immediate conclusion of a commercial transaction, but they could lead to the creation of one or several *witnesses*, people who might later *attest* to the properties on display (those of fuzzy products and fuzzy logic). In this sense, these manifestations stemmed partly from an enterprise dedicated to manufacturing witnesses.[33] Moreover, some researchers in fuzzy logic judged that the "rise" of fuzzy logic depended heavily on the development of fuzzy products, for these made it possible to bring the properties of fuzzy logic to light. According to them, such mediations were capable of ensuring the spread of all the ontological forms of the object and of arousing a growing interest in research into the "formal" properties of fuzzy logic.[34]

[31] See in particular Gooding, Pinch, and Schaffer, *The Uses of Experiment.*

[32] Such gropings are themselves quite remote from the "public experiments" described by historians of the experimental sciences.

[33] On this subject, see also Shapin and Schaffer, *Leviathan and the Air-Pump.* On the conditions and the social role of witnessing in other contexts, see Renaud Dulong, *Le témoin oculaire: Les conditions sociales de l'attestation personnelle* (Paris: Éditions de l'École des Hautes Études en Sciences Sociales, 1998).

[34] For more advanced results in the sociology of demos, see Claude Rosental, *Les capitalistes de la science: Enquête sur les démonstrateurs de la Silicon Valley et de la NASA* (Paris: CNRS Éditions, 2007); Rosental, "Making Science and Technology Results Public: A Sociology of Demos," in *Making Things Public: Atmospheres of Democracy,* ed. Bruno Latour and Peter Weibel (Cambridge: MIT Press, 2005), pp. 346–49; Rosental, "Fuzzyfying the World: Social Practices of Showing the Properties of Fuzzy Logic," in *Growing Explanations: Historical Perspectives on Recent Science,* ed. M. Norton Wise (Durham, N.C.: Duke University Press, 2004), pp. 279–319; Rosental, "De la démo-cratie en Amérique: Formes actuelles de la démonstration en intelligence artificielle," *Actes de la recherche en sciences sociales* 141–42 (Mar. 2002): 110–20; Rosental, "Histoire de la logique floue."

In the room used for demos at FUZCOL, there were also stands devoted to the promotion of journals. Copies were available, along with subscription forms. The journals on display focused on artificial intelligence, fuzzy logic, and uncertainty theories.

A great many conference participants visited this large room, especially during the morning and afternoon breaks (9:30–10:00 a.m. and 3:00–3:30 p.m.), when no talks were scheduled. Snacks were served, so there was a constant parade of waiters around a central buffet, which reinforced the room's power of attraction. The room was continually occupied by participants engaged in dialogue among themselves or with the de-monstrators at the stands, often in small groups. The room thus constituted not only an effective space for de-monstration, but also a prime locus for conviviality.

A very large hall giving access to the set of lecture rooms completed the space devoted to interactions. The actors stopped there often to engage in dialogue, forming small groups at the center of the room around small tables or in the bar area. A reception booth was usually open in this space; this was where the participants picked up the conference program and proceedings as well as their badges on the first day. The badges indicated the name of the participant and that of his or her home institution; a badge was a *marker* constituting a de-monstrative tool in its own right. In some cases, the actors' reputations were associated more closely with their names than with their faces; badge-wearing thus facilitated individual strategies of approach.

The hall also included *display stands* on which a great number of advertising materials had been placed. These documents called attention to technical journals, books, videocassettes, training programs, market study results, conferences, software, electronic components, and a variety of services, all related to the fields of fuzzy logic, computer science, artificial intelligence, and engineering in general. However, unlike the materials in the demo room, these texts, which in some cases conveyed de-monstrations of the properties of fuzzy logic, were not accompanied by the presence of salespeople-de-monstrators. The texts had a different advantage: they were placed right at the entrance to the room, in a passageway that all participants had to traverse.

Cafeterias and restaurants belonging to the vast hotel complex completed a setup that was highly favorable to interactions. The great majority of the participants gathered in these sites to eat in small groups during the lunch break from 11:00 a.m. to noon. In addition, a reception was scheduled from 6:00 to 8:00 p.m. the evening before the colloquium opened, and a banquet was held from 7:00 to 9:00 p.m. on the second day. The fact that a large number of foreign participants were housed on site led to a large measure of integration for these individuals; some of them never left the premises.

Thus important resources had been brought together to constitute a space entirely devoted to de-monstrations and interactions. It must nevertheless be noted that the logic of this concentrating arrangement clashed with competing dynamics. For some participants, visiting the surroundings of the conference represented an opportunity for a touristic adventure. The choice of a place with a pleasant climate for the colloquium favored such projects. As with many other conferences, the site had been chosen with this sort of attraction in mind, as a way of appealing to a maximum number of attendees.

Various forms of academic tourism thus came to light during the colloquium. These had been encouraged by brochures mailed in advance to the participants offering numerous excursions. Under these conditions, and in the face of the series of sometimes repetitive talks, the development of strategies of disinvestment could be observed. To the extent that I remained on the conference site, some participants addressed me with questions that were touristic in nature, before disappearing, provisionally or permanently. Others (usually Occidentals) made joking remarks about the collective disappearances and reappearances of participants from one Japanese research laboratory or another. On the whole, the conference site was increasingly deserted.

If I was able to gather information during FUZCOL, nevertheless, it was because I could circulate freely in the spaces dedicated to exchange. Since I had registered for the conference and was equipped with a badge, I could come and go at will, attend any talk, observe the work of the de-monstrators in the demo room, or carry on interviews anywhere—in the demo room, in the great hall, or in the hotel cafeterias. I was able to meet a number of actors in this way.

At the end of the conference, I was given a list with the names and addresses of the 519 participants. The consensus among the actors I questioned was that all the "great names" in fuzzy logic were present. The participants came from all over the world, but mainly from Japan,[35] the

[35] I was able to question several Westerners about the reasons for the strong participation of Japanese colleagues in FUZCOL. The explanations offered were for the most part cultural in nature, similar to the ones found on the electronic forum. The "pragmatism" of the Japanese, seen as "quick to pick up methods" that work "without too much self-questioning," was regularly invoked to explain the "inroads" of fuzzy logic in Japan. Some asserted that Japan was "cruelly" lacking in engineers, even in priority sectors such as research or industry, the latter having gotten a late start in computerization. According to these informants, the refinement of fuzzy controllers could be done by technicians rather than engineers, which, given the scarcity of a qualified workforce, represented a windfall for the Japanese. Moreover, some of my interlocutors were prepared to think that such a transfer of competences, made possible by fuzzy logic, could lead to an organizational revolution in business enterprises. In the same vein, finally, a young French engineer working in a major Japanese laboratory suggested that one of the reasons for the success of research

United States, and Western Europe. The conference brought together academics, researchers working for research organizations or businesses, engineers, and industry executives. As we shall see, the diversity of skills, practices, and industrial sectors represented was quite considerable. When I asked individuals why they were participating in the conference, the answers fell into several categories.

Some said that they wanted to present their own work, to seek advice or "solutions" to their "problems," to observe the "state of the art" in their particular field, or to get information about the directions research was taking in the short term (results that had not yet been published). Others mentioned their desire to establish contacts, to define future common projects (various types of collaboration, including the organization of future conferences), or to meet colleagues from geographically widespread institutions. This latter aspect took on particular importance for the American researchers. For them, a conference of this sort represented a regular mechanism for interaction that compensated for the effects of significant decentralization in the country. These factors only enhanced the intensity of the exchanges.

The industry executives I met asserted for their part that by registering for FUZCOL they had been seeking to glean information that might guide them in their choice of investments. Such was the case of the American in charge of water distribution in his state. As an illustration, let us look at this example in more detail.

Just as I finished interviewing someone at a stand in the demo room, this American came up to me and spontaneously introduced himself. He explained that he had wanted to participate in FUZCOL because he was soon going to have to upgrade the expert system that regulated the water flow and level in water storage pools for which he was responsible.

He considered the system in place obsolete and wondered what replacement technology would be most "effective." He asked what I thought about this. He also wanted to know whether I thought company X was "better" than company Y. Both companies were represented in the demo room, and my interlocutor expressed a slight preference for the first one because it had branches in the United States, whereas the second one was based primarily in Germany. He wondered, furthermore, whether fuzzy technology was "mature" enough to govern an operation as complex as the one he managed. He also wondered about the compatibility of the system that he might order in relation to the components of the equipment that he wanted to keep. He indicated that he was disappointed overall not to have found a "ready-made solution" to his problems.

---

on fuzzy controllers in Japan stemmed from the fact that the Japanese in general, like the researchers in his own lab, were "no good at math."

To deal with the heterogeneity of the expectations, competencies, and fields of the participants, the conference organizers had put together an impressive program. They had sought to make it as "representative" as possible of this diversity, so that every participant would find something corresponding to his or her own particular interests. The way the papers were selected and the arrangement of the communications helped make this conference as a whole a de-monstrative event, allowing virtually all the research being done in fuzzy logic to be put on display.

Beyond the plenary session, as we have seen, talks were given in three or four parallel sessions held in different rooms. In an interview, David Belmont explained how he had divided the papers into three distinct groups, with the help of his colleague Thomas Berger.[36] The communications devoted to fuzzy and neurofuzzy[37] controllers were assigned to a first group. The presentations that seemed to involve a significant amount of math constituted a second group. "All the rest" were put together in a third series of talks.

Belmont justified these groupings by asserting that the papers devoted to neurofuzzy controllers, the development of electronic components, functional equations, and the applications of fuzzy logic in civil engineering did not interest the same people. He specified jokingly that he had tried to optimize the distribution of papers by using a fuzzy algorithm. This work of thematic construction is a good illustration of the program committee's concern with creating an arrangement that could bring together a large number of participants involved in a variety of fields that were often mutually impenetrable,[38] while appealing to the participants' respective interests.[39]

However, other statements made by Belmont brought out a second concern shared by the program committee presidents: they wanted to be able to offer as complete a panorama as possible of the applications of fuzzy

[36] Let us recall that these men were copresidents of the program committee.

[37] The considerable number of papers devoted to neurofuzzy controls corresponded to a form of sociotechnological alliance developed in the early 1990s between researchers in fuzzy logic and specialists in neural networks, in the face of the partisans of "classical" artificial intelligence.

[38] We should note that Latour's thesis according to which "formalisms" are characterized by the fact that they allow very diverse activities to be coordinated and, in certain cases, controlled from a single place accounts very well for the possibility of such a gathering. See Latour, *Science in Action*, pp. 378–404.

[39] The notion of interest invoked by the actors resembles the one developed in the context of the translation theory presented by Bruno Latour in *The Pasteurization of France*, trans. Alan Sheridan and John Law (Cambridge: Harvard University Press, 1988). See also Michel Callon, "Some Elements of a Sociology of Translation: Domestication of the Scallops and the Fishermen of St Brieuc Bay," in *Power, Action and Belief: A New Sociology of Knowledge?* ed. John Law (London: Routledge and Kegan Paul, 1986), pp. 196–233.

logic, in order to *show* everyone that fuzzy logic was "universal," in the sense that it could be used in (almost) all areas. FUZCOL was one of the first major conferences devoted to fuzzy logic organized by USENG. Thus, for those who were not yet convinced, the organizers wanted to be able to display the existence, scope, and value of research in fuzzy logic in a large number of industrial sectors. This de-monstrative concern was aimed simultaneously at the journalists, industrialists, engineers, academics, and researchers who had come together for the occasion.

Thus the desire on the part of the program committee heads to satisfy the diversified and not always shared interest in fuzzy logic on the part of the participants went hand in hand with the project of highlighting the "universal" properties (in the sense of "universal" defined above) of that logic for the entire set of *spectators*. Among other things, the process entailed manufacturing *witnesses* to the properties of fuzzy logic who would be able to *attest* to those properties in the future.

This project was already quite apparent in the calls for papers that encouraged authors to contribute on the following themes: basic concepts and tools, artificial intelligence, fuzzy neural networks,[40] pattern recognition, control systems, decision analysis, optimization, data and information processing, and fuzzy hardware. During an interview, Belmont declared that it was important for "theoreticians" of fuzzy logic and mathematicians to be present along with the engineers, so they could show the latter the state of advancement of the "theoretical" research being done in this field.

The introduction to the conference program drafted by David Belmont and Thomas Berger specified, nevertheless, that the papers chosen did not offer as exhaustive a panorama of research on fuzzy logic as they had hoped. One of the reasons they gave for this had to do with the process by which the articles were evaluated. Belmont and Berger had had to put the program together after this process had been completed, starting from the papers retained by the evaluators. According to them, the reviewers privileged work being done in the fields they knew best, so that research stemming from sectors that were not yet well developed was less well represented than it "should" have been.

These reflections lead to questions about the precise nature of the process of evaluating articles. To what tests had the articles been subjected? When I asked David Belmont about this during the conference, he explained how the review process had been organized.

Belmont and Berger had invited sixteen researchers to form an executive program committee. Each member was responsible for finding reviewers for the articles he or she was expecting to receive, normally two

---

[40] These involved hybrid "methods," using both fuzzy logic and neural networks.

for each text submitted; the texts were thus divided into sixteen subsets. At the same time, eighteen articles had to be selected for publication in a special issue on fuzzy logic planned by a journal devoted to the theme of neural networks under the auspices of USENG.

According to Belmont, this operation had taken two months. The written conference program indicated the names of the 170 reviewers who had been asked to help. Fifteen percent of the articles, the "best" ones, had been submitted to a single reviewer; 80 percent had been evaluated by two reviewers; the "remaining 2–3 percent [sic]" had been sent to three reviewers. Belmont explained that a third reviewer was sought when it was a matter of "shoring up" a rejection. Once again, we see that the work of those judging a proof—in this instance, the reviewers and editors—was first and foremost a "de-monstrative" activity; it consisted in de-monstrating the well-foundedness of a particular point of view, or even justifying it.

This collective effort and the considerable marshaling of resources and division of labor it entailed resulted in the retention of 185 papers from the 270 that had reached the organizers. But we still need to know just how the selection process was carried out. How did certain articles end up being labeled "bad," "good," or "better" than the others?

An interview with one of the authors of the 185 papers brought partial responses to these questions. This author had had a good deal of experience as a reviewer and conference organizer. During the interview, he addressed his remarks simultaneously to me and to a young "novice" who was attending the talks without presenting a paper himself, and who was asking similar questions. This young man acknowledged that he was surprised to see that a large number of the talks were redundant, presenting the same "basic concepts" of fuzzy logic. He wondered out loud, not without a certain irony, about the "care" that had been taken in the selection of papers.

As an answer, our common interlocutor, whom I shall call Kyle Kelley, underlined the specific objectives of the conference organizers. According to him, the primary goal was to select "good" articles that were "representative" of the research being done in fuzzy logic. The redundant character of the introductions to these papers constituted an accessory problem in their eyes. Kelley offered other details on the way the process of evaluating texts took place.

Each of the sixteen members of the program committee was responsible for selecting a set of communications corresponding to a specific "theme." A quota of articles to be retained had been established. Each person responsible for a theme entrusted the evaluation of the texts to reviewers of his or her choice. Depending on the opinions expressed by these reviewers, each member of the committee was supposed to establish a ranking of the papers he or she had received.

According to Kelley, this work was "anything but self-evident." The articles fell into three broad groups: the "bad" ones and the "very good" ones, which represented a minority of the texts and for which the task of selection proved to be "easy," plus a "morass" consisting of the large majority of the papers, for which the decisions to accept or reject were deemed "delicate." As a general rule, the "very good" articles were written by "well-known" authors. The rejection of the "bad" ones was not a source of anguish for the program committee members; in each one, some element was missing or false, or the article was shoddily constructed. In these cases, rejection was "easy to justify" to the authors.

In contrast, according to Kelley, the "morass" consisted of articles that could be either accepted or rejected. Everything depended, then, on the "quotas" and "goals" that had been set. A member of the program committee who was not convinced that an article contributed "something new" could nevertheless decide that it was "not necessarily uninteresting" for a first conference, in which one does not "necessarily" expect to find "new ideas," but rather works that are "representative" of the various fields of interest. And "from there on," everything depended on the individual reviewers' representations of the nature of the fields of research involved. The reviewers' opinions influenced the decisions without being absolutely determining. The reviewers were judged "more or less competent" to formulate opinions on the papers they received. When a member of the program committee sent a text to a given reviewer, it was with the assumption that the evaluator in question would be able to "say something," even if his or her competence in the field was not "perfectly" adequate.

Kelley indicated that he had served as a reviewer on many occasions. According to him, the modes of evaluation adopted varied considerably from one reviewer to another. He saw himself as fairly decisive in this role. For him, an article whose topic and main argument were not explicit was "not serious" and "off to a bad start." So long as a text was "comprehensible," the opinion he could formulate was not entirely based on "purely rational" considerations, he claimed. There were "things" about which "one was more or less enthusiastic," and he said that he himself was more inclined, "like more or less everybody," to "push for things he liked"—an attitude he deemed "pretty human"—while staying nevertheless "within limits" where "it isn't just that."

He characterized the selection process that had been followed for FUZ-COL as "reasonable" to the extent that the rejection rate was around 50 percent. According to him, this proportion should "normally" lead, "on the average," to a range of "interesting" and "reasonable" articles that "have something to say." He felt that a more rigorous selection would not have made much sense, for it would not have been as representative of the research being undertaken in fuzzy logic.

As a comparison, Kelley evoked the case of a conference in which he had sought to participate. The rejection rate for papers had reached 90 percent, and his proposal had been rejected. According to him, it was "ridiculous" to accept so few papers, and it made the process a sort of "lottery." He thought the organizers would have been quite unable to justify why a given paper had been selected rather than some other, if that question had been raised.

His text had been sent back to him with the reviewers' annotations. He saw these as focusing merely on details—he was reproached, for example, for putting "that" instead of "what," or for not dealing in one passage of his text with the question of the way a device operated. His paper had been characterized as an excellent article for a journal, a comment that he did not deem "misplaced" to the extent that the text developed a general problematics rather than presenting a "somewhat partial" result, as was the custom, according to him, in conferences. He thought that this comment had constituted a simple decision-making criterion, during the program committee meeting, as well as a good argument for the rapid rejection of his text.

But my interlocutor was persuaded that a "ferocious" hostility to fuzzy logic had in fact motivated the decision to reject his paper. To his knowledge, the only paper by a researcher in fuzzy logic that had been accepted at a major conference devoted to his own field of research (having to do with database management) had been in 1976. According to him, in this milieu it sufficed for the reviewers to see the term "fuzzy" for them to try to "debunk" the text. And he thought that a reviewer who really wanted to "debunk" a paper, any paper at all, was always in a position to succeed. He could assert, for example, that what was presented was "of no use." He reported that a number of his colleagues doing research in fuzzy logic had already faced this experience.

He affirmed that, in contrast, some articles were published without being submitted to reviewers. It sometimes happened that articles were retained for publication without being read. In the context of a conference that he was helping to organize, he had thus gotten articles accepted with no prior evaluation. These were cases in which texts had been solicited from authors he knew well, with whom he had reached an understanding in advance about the themes to be addressed. In contrast, he had had his own article evaluated, and it had been accepted without comment. This led him to say that at all events it would have been difficult for the other organizers to reject his text.

This testimony sheds new light on the process of selecting papers for FUZCOL and other conferences. The decision to publish a text (or not) depends on whether (or not) it succeeds in passing various tests. The work of selecting articles is not an abstract operation in itself. It proceeds from a series of material tasks: the circulation of articles, a cascade of readings, acts of writing (reviewers' annotations of texts, drafting of letters and reports). The process involves numerous interactions between program committee members, reviewers, and sometimes even authors.

In the case we are following, the ultimate fate of each article appeared to depend quite directly on the way the group mobilized in the endeavor worked together, and in particular on the fact that the article was assigned to a given member of the program committee, transmitted to certain reviewers, and confronted with a specific rejection rate. In this sense, while some groups indeed "supported" the publication of Elkan's de-monstration or opposed it in the FUZLOG context, this was largely because the selection of articles stemmed neither from a nonmaterial dynamics nor from an individual undertaking.

Moreover, in light of Kyle Kelley's testimony, we can see what a long list of elements came into play in the decision-making process. Kelley did not evoke "universal" criteria of "correctness" applicable to the proofs that would have made it possible to select articles "automatically." To be sure, he did allude to results that were deemed erroneous. But when he sought to account for the process of evaluating papers, he spoke mainly about uncertainties, hesitations, and awkward moments; he referred to authors' reputations, modes of appreciation that varied from reviewer to reviewer, relative competencies, irrationality, degrees of rigor, indeterminate boundaries that must not be crossed, personal representations and tastes, averages, arbitrariness, ill will, types of publications, representativeness, and prior understandings. The relatively vague and opaque character of the terms used to describe the way the reviewers assessed the texts ("shoddily constructed," "reasonable," or "not very interesting" papers) or to describe their own work ("fairly decisive," "more or less enthusiastic," "debunking") bears witness as such to the complexity of the process for selecting articles and to the diversity of elements that came into play in the decisions reached.

This testimony gives a more precise meaning to, and extends the scope of, the descriptions by Davis and Hersh that we examined earlier. Relying on their own experience, those two mathematicians asserted that the evaluation of the correctness, the value, or the problematic character of a proof, a result, or an article, stemmed from local skills proper to a subdiscipline, skills often inaccessible to people who are not part of the research field in question. For Davis and Hersh, the author's reputation and the way the editors of a journal represented the research field covered by the journal were also factors influencing the decisions to accept or reject a paper.[41]

In the case under study, the way the reviewers and program committee members looked at the authors' "reputations," and even more what they saw as *representative* research in a given field (here, fuzzy logic), appeared to play a role in the selection of articles. In addition, the judgments brought to bear on the correctness of proofs, on what might be problematic or "interesting" in a text, were considered above all as being in the purview of researchers in fuzzy logic, and were in fact made by researchers in this domain.

Furthermore, by stressing the difficulties encountered in the distribution of articles to reviewers, the actors highlighted the need for very specialized competence in order to carry out evaluations. One might have supposed that a sort of "neutrality of formalism" in relation to the objects would have allowed researchers in fuzzy logic, starting from a mastery of canonical symbolic manipulations, to grasp the activities of the field as a whole. As we have seen here, this was far from the case. Moreover, the specificity of the skills required was such that, despite the mobilization of a considerable number of reviewers (170 for 270 articles), expressions of dissatisfaction came up again and again.

However, Kyle Kelley's testimony introduced entirely new elements with respect to the descriptions offered by Davis and Hersh. Kelley brought to light the complex and uncertain character of the work of evaluating texts and the forms of creativity this work required (hence the "anguish" of reviewers and program committee members in the face of a "morass" of texts that were difficult to categorize). Not only did the program committee members and reviewers have no closed and definitive list of criteria or methods that it would have sufficed to apply in order to select articles "automatically," but an essential dimension of their activity consisted in *searching*, often on a case-by-case basis, for some elements of appreciation that would allow them to make a judgment or decision and de-monstrate its well-foundedness (as illustrated for example by the notation "excellent journal article").

In consequence, the work of those who judged proofs appears once again to belong first and foremost to the category of de-monstrative activity. When Kelley described how the choice of reviewers by program committee members was made in relation to their presumed or tested ability to "say something," he brought into relief the search for a capacity, judged primordial, to produce an opinion and de-monstrate its rational character. In the light of Kelley's testimony, it becomes apparent that one of the qualities required to play the role of reviewer or program

---

[41] See Davis and Hersh, "Rhetoric and Mathematics."

committee member consists in being able to develop and arbitrate between several *registers* of appreciation—for example, being able to balance a text's "lack of originality" against the degree to which it is "representative" of a research field, in order to shore up a decision to accept or reject a given article.

Thus it would seem that the work of evaluating articles stems in one important way from the reviewer's capacity to escape from "inexpressibility," the vocabulary of taste, or the observation of the irrationality of the choices (phenomena to which Kelley referred abundantly), in order to be able to win the assent of the potential addressees of his or her comments (program committee members, or editors, and sometimes also authors, who may demand to know why the paper was rejected). Under these conditions, it seems possible to characterize a "bad" article in the following terms: it is a text whose rejection can be easily justified.

Moreover, while Davis and Hersh maintained that a given article could be the object of either a favorable or an unfavorable judgment depending on the reviewers to whom it was submitted (a claim that the story of Elkan's text illustrates perfectly), Kelley's testimony sheds a little more light on the subject. It emphasizes not only the variable nature of reviewers' evaluation practices, but also the considerable margins for maneuver reviewers enjoy as they carry out their task. We note this when Kelley mentions the opportunity a reviewer has to "push the things he likes," to "debunk an article when he really wants to," or to say that a theory or method "is of no use."

By relaying John Carpenter's denunciatory testimony, Kyle Kelley's account has brought a conflict between two groups to center stage.[42] It shows how this antagonism can unfold on the terrain of the choice of publications, by relying on the maneuvering room that is available to reviewers or editors as they formulate their opinions. Not only must we take into account the existence of potential conflicts between actors, then, but also the whole set of practices governing the selection of articles adopted in the context of conferences, if we want to understand why Elkan's de-monstration was accepted by the AAAI'93 committee and rejected by that of FUZLOG.

Finally, the considerable number of mediations that intervene in the selection of conference papers explains the opacity of such a process for a young researcher (such as the "novice" to whom Kelley was also speaking during our interview). Laying out these mediations here does not stem from any intent on my part to expose something that had been concealed. For experienced researchers, the givens of this process were neither "dis-

---

[42] In Davis and Hersh's terminology, these groups would certainly be characterized as "subdisciplines."

simulated" nor "surprising." The fact that certain works of epistemology do not take them as an object and attempt rather to discern "methods" and "criteria" for evaluating the "correctness" of proofs does not signify that these elements are "hidden" by the actors.[43] They are all the less hidden, moreover, in that they constitute an object of debate for practitioners.[44] These elements must in any case be taken into account, as I hope to have shown, in any attempt to discover what tests Elkan's demonstration had to pass in order to be published, and how his text could have been both rejected at FUZLOG and accepted at AAAI'93.

[43] See Jerome R. Ravetz, *Scientific Knowledge and Its Social Problems* (Oxford: Clarendon Press, 1971).

[44] In particular, we observe no agreement here about what would constitute the "content" and the "context" of the logical exercise, about what it is essential to "find" or to "verify" in an article that includes one or several logical theorems.

Chapter 6

FEDERATING A COUNTER-DE-MONSTRATION
OR PRODUCING HAND-TAILORED RESPONSES

N OW THAT THE development and publication phases of Elkan's
proof have been analyzed, we need to return to the unfolding of
the controversy that began in the summer of 1993. The analyses
completed so far raise several questions. First of all, what set of undertakings accompanied the interventions on comp.ai.fuzzy? Did certain protagonists succeed, in a way still to be determined, in federating collective
actions and imposing dominant viewpoints? What role did the exchanges
on the electronic forum actually play in the general dynamics of the debates, and precisely how were these debates mediated? In seeking to answer these questions, let us begin by examining some of the actions undertaken by researchers in fuzzy logic during the period under study, looking
in particular at the way the actions were coordinated.

Producing More Stable and Visible Responses, in Limited Number

The study of the debates over Elkan's proof on the electronic forum gave
rise to an initial impression of cacophony. But it gradually became clear
that certain messages had acquired higher visibility than others and had
served as references for many participants. Several mechanisms drew attention to the interventions of David Belmont, Thomas Berger, and John
Carpenter. The tone adopted by the latter in particular, with its federating
accents, made his messages stand out. It thus seems worthwhile to begin
by looking at the actions John Carpenter undertook alongside his interventions on the electronic forum.

Adopting the role of spokesperson for researchers in fuzzy logic, Carpenter had sent a letter of protest to the organizers of the AAAI'93 conference. The account of this he gave me in 1994 must be viewed as a *counter-
de-monstrative* undertaking in itself, since it made me a witness to an
instance of unfair behavior. The positions Carpenter developed were very
close to those displayed on the electronic forum. During our interview, he
found it crucial to offer an effective counter-de-monstration to challenge

Elkan's theses, given the visibility they had been granted, the discredit they risked bringing to fuzzy logic, and the danger they represented for research funding in the field. In this context, Carpenter mentioned additional projects for countering Elkan's paper.

He indicated that he was in the process of writing an abridged version of the message he had sent to comp.ai.fuzzy. This text was intended for publication in 1994 in a major journal in the field of artificial intelligence. Carpenter explained that this journal had granted him the right of response but that it imposed strict constraints on the length of his article.

He planned to intervene in collaboration with two researchers in fuzzy logic, Peter Smith and another actor whom I shall call Frank Hughes. Each of their texts had to be limited to 1,500 words, and the three were to contribute complementary responses to Elkan's critiques. Carpenter would focus on Elkan's proof; Smith would challenge Elkan's theses concerning fuzzy controllers; Hughes would offer objections to Elkan's claims regarding expert systems. Elkan's reactions to these criticisms would be published in the same issue.

As Carpenter saw it, this project was all the more important in that to his knowledge no refutation of Elkan's theses had been published up to that point. The only debates he had observed by early 1994—nearly six months after the publication of Elkan's article in the AAAI'93 proceedings—had taken place exclusively on the comp.ai.fuzzy forum.

In the light of this testimony, we can see how John Carpenter was reinforcing his position as spokesperson, throughout 1994, by intervening on a new stage. While he had already distinguished himself from the other participants in the discussions on the electronic forum, the gain in visibility that this new publication might represent was much greater. Here we can see how a limited group of researchers in fuzzy logic managed to occupy center stage after a large number of actors had formulated their objections. This point warrants elaboration, for it introduces elements that are essential for grasping both the evolution of the debates and the conditions under which the de-monstrative activity took place.

While the economy of the electronic forum's operation initially allowed the display of multiple contradictory de-monstrations that benefited from virtually identical visibility, the pursuit of the exchanges in specialized journals marked a turning point in the controversy. Certain participants disappeared from the debates and remained silent thereafter, whereas others (such as Carpenter) managed to keep the floor and increase their own ability to be heard.

Like Carpenter, I myself was unable to locate any debates over Elkan's article before early 1994 except on comp.ai.fuzzy—hence the importance of this forum and the usefulness of a detailed analysis of the exchanges it fostered. However, to my knowledge, very few participants in the comp.ai.fuzzy discussions expressed their criticism in journal articles later on (during the first half of 1994). The group challenging Elkan was limited to a small number of individuals, including John Carpenter, David Belmont, and Thomas Berger.

The resources available to these last three individuals even before the beginning of the controversy were of course not irrelevant to the situation. These men all possessed a powerful ability to promote their own viewpoints and to offer structured responses playing on numerous registers. They had each long been committed to and visibly invested in the mission of representing and defending fuzzy logic. The reputations they had acquired and the scope of the relational systems they had been able to forge were elements to which most of the other contributors to the electronic forum clearly lacked access. In this context, the fact that Carpenter, Belmont, and Berger had taken over center stage was hardly surprising.

In 1994, the multitude of viewpoints formulated by researchers in fuzzy logic began to give way to a counterargument that was *structured* by a few confirmed representatives of the field. Some of them, for example Peter Smith and Frank Hughes, never expressed themselves on comp.ai .fuzzy. Still, like Carpenter, prior to the publication of Elkan's article these two had served as spokespersons for fuzzy logic. Smith and Hughes had also headed conference program committees and had contributed to continuing education training seminars, often working together.

As we have seen, one of the most striking features of the proposed contributions of fuzzy logic researchers to journals, starting in 1994, was their highly structured character. This phenomenon manifested itself at one level through the drafting of particularly well-constructed counter-de-monstrative texts, comparable to John Carpenter's intervention on comp.ai.fuzzy.

At another level, the *stability* of the positions taken, we note the great consistency of Carpenter's judgments in the context of comp.ai.fuzzy, in the interview he granted me, and in the publication he was contemplating in a journal devoted to artificial intelligence. In the face of the multitude of viewpoints expressed on the electronic forum, John Carpenter's demonstrative undertaking was largely grounded in the reiteration of an unchanging critique.

The structured character of the projects for intervening in journals can be situated at still another level: in the *federation* of the responses made to Elkan's article. Carpenter coordinated his actions with those of his two colleagues. By being careful not to produce contradictory or redundant objects like those found on the electronic forum, these researchers in fuzzy logic thus developed a more "effective" counterargument.

Several factors contributed to the formulation of a unified response to Elkan's de-monstration by fuzzy logic specialists starting in 1994. In particular, few authors were in a position to express their views in journals. The time and skills needed to craft a response supported by arguments that respected powerful constraints on form and content might well have been dissuasive for some. In any case, editorial restrictions limited the number of articles that could be accepted. Coordination among the few researchers who were able to express their own points of view was one answer to such constraints.

In other words, while the positions of some researchers in fuzzy logic stood out and became central, this was not because all the participants in the electronic forum were convinced of the correctness of these positions. Various elements made it possible for certain participants to remain in the controversy and to federate a counter-de-monstrative action, while others disappeared from the scene of exchanges. Let us note in passing that what was going on here had little in common with a model of rationality according to which the evolution of any debate in logic would simply be determined or characterized by the nature of the "arguments" produced.

We must also note the publication by several major figures in fuzzy logic of a common response to Elkan's de-monstration in a journal devoted to artificial intelligence in the spring of 1994. This article was cosigned by eight authors, including Didier Dubois and Henri Prade.[1] It revealed a demonstrative undertaking that privileged the formulation of a stable and univocal critique (resembling John Carpenter's), one that evolved little over time or from one forum to the next. Once again, then, the structured character of the responses of the major actors in fuzzy logic was apparent. The actors' ability to offer themselves as spokespersons for fuzzy logic in various forums in artificial intelligence[2] was an important factor behind the consolidation of their response.

The controversy thus underwent a major evolution beginning in early 1994. The list of its participants, its pace, and the dynamics of the exchanges were all profoundly modified. Among the researchers in fuzzy logic who intervened on the electronic forum, only those who managed to publish remained in the game. The debates now brought together only the leading figures in fuzzy logic. Comp.ai.fuzzy was no longer the essential site for debate that it had been up to the end of 1993. The publication

[1] Hamid R. Berenji, James C. Bezdek, Piero P. Bonissone, Didier Dubois, Rudolf Kruse, Henri Prade, Philippe Smets, and Ronald R. Yager, "A Reply to the Paradoxical Success of Fuzzy Logic," *AI Magazine*, Spring 1994, pp. 6–8.

[2] See also the only response to Elkan's article formulated by researchers in fuzzy logic in the context of AAAI'94: Didier Dubois and Henri Prade, "Can We Enforce Full Compositionality in Uncertainty Calculi?" *Proceedings of the 1994 National Conference on Artificial Intelligence* (Cambridge: American Association for Artificial Intelligence, MIT Press, 1994), 149–54.

of objections to Elkan's thesis in some artificial intelligence journals became a central element in the counterthrust organized by the leading figures in fuzzy logic. The pace of the controversy was profoundly modified by this shift, because the time needed for publication imposed a less frenetic rhythm and introduced some spacing between de-monstrations and counter-de-monstrations. The various interventions were now separated by months, a duration that became the time scale adopted by the protagonists still "in the running." This time scale structured the way the latter projected themselves into the future. The initial contradictory critiques were succeeded by a period in which de-monstrative actions were highly coordinated, and responses were more univocal.

Moreover, the increasing constraints that weighed upon the authors' interventions included editorial restrictions on the length of texts. In this connection, we note once again how material constraints (such as the instruction "no more than 1,500 words") could determine the actors' activity and their productivity in a fundamental way. We may also appreciate the major changes brought about by recourse to the electronic forum, as compared to more traditional media, in the progress of such debates. The importance of the material mediations of the de-monstrations, the diversity of the undertakings, and the coordinated character of the interventions also highlight the distance that separated the unfolding of the controversy from a pure "exchange of arguments." A simple study of the argumentation, or a transposition of the debates into a classroom context, analogous to the one Lakatos undertook in order to analyze the controversy over Euler's theorem,[3] would be quite unable to account for such essential elements of the de-monstrative activity.

By bringing to light the important role played by John Carpenter and a few others in the pursuit of the debates starting in 1994, these descriptions justify a posteriori the particular attention we have paid to their actions. However, the reader may well wonder at this stage why our picture does not feature a well-known researcher in fuzzy logic such as Bart Kosko. After all, Kosko was the author of the text on which Elkan's article was based.

This eclipse is not the result of the use of pseudonyms in the narrative. As it happened, I found no trace of any intervention by Kosko in the debates. This somewhat puzzling discretion warrants consideration.

It could be that Kosko's silence was a consequence of the fact that various researchers in fuzzy logic had rejected certain of his arguments. It seems quite unlikely that agreement with Elkan's thesis caused Kosko's silence. By remaining relatively undemonstrative (or rather un-de-monstrative), Kosko could avoid getting involved in a tangled and potentially

[3] Lakatos, *Proofs and Refutations*.

bitter debate with Elkan or with researchers in fuzzy logic who did not share some of his theses. Several elements point up the existence of critical attitudes toward Kosko's positions among several actors in fuzzy logic. A close look at these elements may offer a better grasp of the origins of the various dynamics of investment and disinvestment that characterized all participants in the controversy.

To begin with, let us recall that the debates on the electronic forum brought to light dissensions among researchers in fuzzy logic as to the nature of their objects. These dissensions were particularly evident with regard to the proculturalist and anticulturalist theses used by the actors to account for the development of that theory. Without naming names, Ruspini denounced the adoption by "certain" individuals, in "certain" publications, of positions stemming from a form of cultural relativism.[4] During his interview with me in early 1994, Carpenter affirmed that such positions were defended by Kosko and some of his partisans. What theses was Kosko propounding, then, and in what ways?

In the article that served as a point of departure for Elkan's text, Kosko proposed a de-monstration of the following claim: fuzzy logic makes it possible to solve the paradox of Russell's barber.[5] He de-monstrated more precisely that this paradox could be resolved by renouncing the principle of the excluded middle and by using a form of logic with several truth values.[6] In the developments following this de-monstration, Kosko pos-

---

[4] On this topic, see also Enrique H. Ruspini, "An Opportunity Missed," *IEEE Spectrum*, June 1993, 11, 12, 15; Dan McNeill and Paul Freiberger, "Out of Focus on Fuzzy Logic," *IEEE Spectrum*, August 1993, p. 6; Enrique H. Ruspini, "Fuzzy Logic Revisited," *IEEE Spectrum*, September 1993, pp. 6, 8; Dan McNeill and Paul Freiberger, "The Authors Respond," *IEEE Spectrum*, September 1993, pp. 6, 8; Pease Porridge, "What's All This Fuzzy Logic Stuff, Anyhow?" *Electronic Design*, May 13, 1993, pp. 77–79.

[5] In his article, Kosko presented this paradox in the following terms: "Russell's barber is a bewhiskered man who lives in a town and shaves a man if and only if he does not shave himself. So who shaves the barber? If he shaves himself, then by definition he does not. But if he does not shave himself, then by definition he does. So he does and he does not –contradiction ('paradox')." See Kosko, "Fuzziness vs. Probability," pp. 219–20.

[6] The solution proposed by Kosko was openly inspired by the one formulated in Gaines, "Precise Past, Fuzzy Future." Immediately after setting forth the paradox, Kosko "solved" it as follows: "Let S be the proposition that the barber shaves himself and not-S that he does not. Then since S implies not-S and not-S implies S, the two propositions are logically equivalent: S ⇔ not-S. Equivalent propositions have the same truth values: $t(S) = t(not\text{-}S) = 1 - t(S)$. Solving for $t(S)$ gives the midpoint point of the truth interval . . . : $t(S) = 1/2$. . . . In bivalent logic both statements S and not-S must have truth value zero or unity. The fuzzy resolution of the paradox only uses the fact that the truth values are equal. . . . The midpoint value 1/2 emerges from the structure of the problem" (Kosko, "Fuzziness vs. Probability," pp. 219–20).

For a detailed analysis of this resolution, see Rosental, "Histoire de la logique floue." It should be noted that other resolutions of "classical" logical paradoxes by fuzzy logic have been proposed in Lotfi A. Zadeh, "Liar's Paradox and Truth-Qualification Principle," Uniited the realistic character of the descriptions of fuzzy logic, along with its historical and cultural properties. He then established a connection between these latter properties and the specific history of the barber's paradox—which, along with other paradoxes, is considered to have marked the end of logicism at the beginning of the twentieth century.[7]

According to Kosko, fuzzy logic constitutes the only realistic description of the universe, for any statement about the universe is in general neither entirely true nor entirely false. As he sees it, this logic escapes the *contradictions* of binary logic and of the principle of the excluded middle, and the term "paradox" (as opposed to "contradiction") is only a euphemism intended to avoid an acknowledgment of failure. In fact, for Kosko, the barber's paradox signals the contradictory character of binary logic and especially of the principle of the excluded middle (according to which a proposition is either true or false).

As a rationalist, the author considers that he has de-monstrated in his article that the resistances to giving up binary logic and the principle of the excluded middle are themselves irrational, and can only be explained by historical and cultural factors. For Kosko the historian, "faith" in binary logic and in the principle of the excluded middle stems from a cultural predilection transmitted in the West by a pedagogical tradition that goes back at least to Aristotle. According to him, fuzzy logic marks a turning point between the end of binary Western thought and the coming predominance of Eastern thought and civilization.

The de-monstration of the appropriateness of fuzzy logic for solving the barber's paradox is thus itself the starting point for a de-monstration that mixes various registers in order to call on the realist, historical, and cultural properties of that logic. Several studies in the sociology of science have shown how scientists can seek to reconstruct the world from their laboratories.[8] We can see here how an academic, through the mediation of a simple article, and starting from the graphic deployment of a logical property, can attempt to reconstruct the history of civilizations and "universal" categories of representation.

---

versity of California, Berkeley, Memorandum no. UCB/ERL M79/34, May 18, 1979. On this subject, see also J. F. Baldwin and N.C.F. Guild, "The Resolution of Two Paradoxes by Approximate Reasoning Using a Fuzzy Logic," *Synthese* 44 (1980): 397–420.

[7] For other resolutions of the barber's paradox and a study of its relative importance in the history of logic, as compared to the role played by other paradoxes (whose similarities and differences are analyzed), see R. M. Sainsbury, *Paradoxes* (Cambridge: Cambridge University Press, 1988), esp. p. 110. For another approach to the barber's paradox (and the liar's paradox), see Jon Barwise and John Etchemendy, *The Liar: An Essay on Truth and Circularity* (Oxford: Oxford University Press, 1987), esp. p. 172.

[8] See especially Bruno Latour, "Give Me a Laboratory and I Will Raise the World," in *Science Observed: Perspectives on the Social Studies of Science*, ed. Karin D. Knorr-Cetina and Michael Mulkay (London: Sage, 1983), pp. 141–70.

However, other mediations took up where this simple text left off, and carried the set of properties thus defined even further. In the early 1990s, Kosko published a book that met with great success and became a major vehicle for the definition of fuzzy logic. *Neural Networks and Fuzzy Systems*,[9] which associated fuzzy logic with the properties we have just examined, was used as a textbook in departments of computer science and engineering in a number of American universities. According to several concordant accounts, this text sold more copies in the United States than all other works on fuzzy logic combined, making Kosko one of the central figures in the field.

There were other important vectors for conveying Kosko's particular definition of fuzzy logic: Kosko himself wrote a book for the general public on the topic;[10] a volume by two other authors had presented Kosko's theses to the same general audience the previous year.[11] Additional resources included journal articles, courses at the university where Kosko was teaching (the University of Southern California, in Los Angeles), and—also in the United States—conferences, continuing education seminars, book signings in bookstores, radio appearances, interviews granted to journalists cited in newspaper articles, advertising campaigns for Kosko's writings in electronic forums, in specialized journals or in newsletters of engineering associations, and so on.[12] In short, a whole set of mediations came into play, mediations not specific to the promotion of logical objects; they could be found in other forms of academic life where practices of writing were essential.

Given the proliferating modes of exhibition of the previously described properties of fuzzy logic and the growing propagation of the corresponding definition of fuzzy logic, some researchers in the field—including John Carpenter—began to form a coalition to promote a competing definition (or definitions) of this object. During an interview in 1994, Carpenter argued that W.V.O. Quine had succeeded in showing that the barber's paradox could not be solved by using a form of logic with several truth values.[13] Carpenter viewed Kosko's translations as incorrect; fuzzy logic did not allow for a solution to the barber's paradox. The historical and cultural properties attributed by Kosko to fuzzy logic were entirely con-

---

[9] Bart Kosko, *Neural Networks and Fuzzy Systems: A Dynamical Systems Approach to Machine Intelligence* (Englewood Cliffs, N.J.: Prentice Hall, 1992).

[10] Bart Kosko, *Fuzzy Thinking: The New Science of Fuzzy Logic* (New York: Hyperion, 1993).

[11] Dan McNeill and Paul Freiberger, *Fuzzy Logic: The Discovery of a Revolutionary Computer Technology, and How It Is Changing Our World* (New York: Simon and Schuster, 1992).

[12] See in particular comp.ai.fuzzy, message no. 1234.

[13] Quine, *Ways of Paradox*.

testable, from Carpenter's standpoint, as much so as the theses about the history of science that led Kosko to denigrate the Aristotelian tradition. For, according to Carpenter, logical systems with several truth values had in fact been developed within that very tradition.[14]

Beyond the criticism that emerged during this interview and that led to the promotion of a different definition of fuzzy logic, Carpenter was engaged in other actions. He had just published a critique of a work for the general public that took up Kosko's themes, and this critique had gotten him involved in a polemic with that work's authors. He was also participating in the preparation of a textbook; in its first chapter, he proposed an exercise that would make it possible to prove that Kosko's solution to the barber's paradox was erroneous. The modalities of this deconstruction (once again in graphic form) consisted in identifying and challenging one of the translations, and in replacing this branch of the proof with a new series of translations leading again to a paradox.[15]

Another example of the many endeavors to which such dissensions gave rise can be found in some of the activities surrounding a major conference

[14] In the course of my study, researchers in fuzzy logic offered various histories of this logic, in connection with their own de-monstrative aims. In a number of cases, the history of fuzzy logic was presented as the history of an "invention" that originated in the mind of L. A. Zadeh. This invention was associated with the publication of a first article on this subject in the mid-1960s (see Zadeh, "Fuzzy Sets"). The narratives were constructed like histories of ideas, and sometimes like histories of events. They were then sprinkled with institutional, technical, and industrial data (dates of creation of laboratories, associations, or journals, and also, for example, dates when conferences were held or dates when products associated with fuzzy logic went on the market). In relation to other historiographic registers, the authors attributed more ancient origins to fuzzy logic, and they inscribed its emergence within the rise of the forms of multivalent logic developed by Lukasiewicz, Russell's philosophy of vagueness, Max Black's theory of vague sets, or even the philosophy of Heraclitus, by way of Plato, Hegel, Marx, Engels, and Knuth. For more details, see Rosental, "Histoire de la logique floue." See also comp.ai.fuzzy, messages no. 40, 124, 1555, 2071; Brian R. Gaines and Ladislav J. Kohout, "The Fuzzy Decade: A Bibliography of Fuzzy Systems and Closely Related Topics," *International Journal of Man-Machine Studies* 9 (1977): 1–68 (this article includes an extensive bibliography on fuzzy logic and related topics through the mid-1970s); Bart Kosko and Satoru Isaka, "Fuzzy Logic," *Scientific American*, July 1993, 76–81; Max Black, "Vagueness: An Exercise in Logical Analysis," *Philosophy of Science* 4, no. 4 (1937): 427–55; *Synthèse technologies de l'information*, vol. 1 (Paris: Aditech, 1989–90). For a systematic analysis of the ways in which the history of an invention may be constructed, see Hélène Mialet, "Le sujet de l'invention," Ph.D. thesis, Paris, Université de Paris I–Sorbonne, 1994.

[15] Once again, the processes of transformations and visual displacements I am describing are different from what Kuhn called "revolutions as changes of world view," if only because I am not bringing paradigm modifications into play. From my perspective, if the worldviews of relatively large groups of actors turn out to be modified, this comes about through the chain of mediations that can intervene *even* at the level of interactions. However, the analysis of the diversity of these mediations cannot be reduced to general theses of cognitive

on fuzzy logic that took place prior to the publication of Elkan's article. The organizers thought that Kosko's theses were erroneous, that they gave a false image of fuzzy logic and threatened its credibility. Thus they deemed it important to place the paper Kosko had proposed in a thematic session that did not touch in any central way on the definition of fuzzy logic. In personal encounters during the conference, they revealed their fears to certain participants, encouraging the latter to bring up the problem and to emphasize the errors Kosko had made in his talks. They insisted that Kosko's theses threatened all research activity in the field, especially in the context of a "cold war" with the proponents of other approaches in artificial intelligence, given the increasingly intense competition for the same sources of funding.

During our interview at FUZCOL, David Belmont shared his concern about the growing visibility of Kosko's theses; he thought that they threatened to provide fodder for criticisms of fuzzy logic and might even discredit it. He thus accurately anticipated the publication of Elkan's article a little more than a year later in the proceedings of AAAI'93. Evoking Kosko's errors orally, as Carpenter had, Belmont nevertheless formulated a different critique. His was more focused on Kosko's thesis according to which fuzzy logic was a generalization of probability theory, something that Belmont contested unequivocally.[16] Moreover, Belmont's critiques (just like the counterthrust he published several months later on the electronic forum) were inscribed in a more algebraic tradition than the one that marked Carpenter's positions. Such critiques corresponded to Belmont's own practice of logic.[17]

By taking into account all these forms of opposition among actors in fuzzy logic over the definition of their object prior to the beginning of the controversy, we are now in a better position to grasp the dynamics of the production of counter-de-monstrations in reaction to Elkan's article. As we have seen, numerous researchers in fuzzy logic had been mobilized against Kosko's theses even before the publication of Elkan's text, and certain coalitions were being formed to limit the diffusion of Kosko's ideas.

---

psychology (in particular Gestalt theory), which constitute the point of departure of Kuhn's analysis. See Kuhn, *Structure of Scientific Revolutions*, pp. 110–34.

[16] For certain researchers in fuzzy logic, the development of this logic was inscribed within the recent and more general emergence of a new field having to do with the treatment of various forms of "imperfect knowledge." According to these researchers, fuzzy logic was part of a set of competing or complementary approaches to probability theory, which had had a monopoly in the field until recently. This set generally encompassed various approaches to uncertainty, vagueness, imprecision, incompleteness, and partial inconsistency, and it included "theories" such as nonmonotonic logics, modal logics, Bayesian and non-Bayesian probability theories, the theory of belief functions, fuzzy sets, possibility theory, the theory of evidence, and belief networks.

[17] See Largeault, *La logique*, pp. 16–24.

The protagonists in the debates, especially on the electronic forum, thus seem even less interchangeable than they might have appeared at first. In retrospect, we can see that the divisions between the defenders of fuzzy logic on the electronic forum did not emerge ex nihilo but proceeded from cleavages that were operating on a larger scale. Taking into account the divergences in viewpoint that were developing among researchers in fuzzy logic, we can finally see why Kosko did not intervene alongside Carpenter or others who signed common responses to Elkan's de-monstration.

By not participating in the debates, Kosko avoided the risk of finding himself caught between the hammer and the anvil, as it were: a response from Elkan on the one hand and those of researchers in fuzzy logic on the other. Let us recall the way an intervention on the electronic forum that took positions close to certain of Kosko's theses had been subject to criticism both from Elkan and from researchers in artificial intelligence and in fuzzy logic (including Ruspini), denouncing culturalist positions that were deemed misplaced.[18]

The way the controversy unfolded is now clearer. Elkan's article can be construed as an extension of Bart Kosko's use of a visible although nonconsensual definition of fuzzy logic. Let us recall that Elkan had formulated his de-monstration with reference to Kosko's article, which he had read earlier, after a first phase in which he had consulted a newspaper article that referred to Kosko.

Moreover, we note that the efforts of certain prominent figures in fuzzy logic in these debates are not limited to the refutation of Elkan's critiques. They also involve a critique of some of Kosko's theses. In other words, the controversy we are studying must also be grasped in the context of phenomena of competition among researchers in fuzzy logic over the definition of their objects.

Limiting the investigation to the study of interventions on the electronic forum would thus not have allowed us a full grasp of their dynamics. We would not have been able to identify some of the resources drawn upon by the participants in that framework; we would not have caught certain implicit references (especially to Kosko's approach). Most of all, we would not have grasped the importance of a past capable of partially determining the actors' interventions—and even their silence.

To be sure, the protagonists' de-monstrations are of the agonistic order, and the tendency of a given definition of fuzzy logic to win out over its competitors depends in particular on the capacity of each participant to take the floor, to put himself or herself more or less in the foreground.

[18] See chapter 4, the section "Contesting a Proof and Defending Logical Properties by Evoking a Cultural Specificity," and in particular Enrique Ruspini, message 816, quoted on p. 140.

But the economy of the interventions stems above all from dynamics that unfold over the long term and from the formation of coalitions among actors; these factors provide more resources for the defense of their viewpoints to certain groups than to others. Moreover, for Elkan, the importance of coalitions and of oppositions among researchers in fuzzy logic, and in particular the importance of the challenge to Kosko's definitions of that logic posed by some major figures in this research field, was a resource in the management of the controversy with which he was confronted—whereas we may recall that before his article was published he did not deem it appropriate to take into account any potential diversity in the definitions of fuzzy logic.

However, the protagonists' ability to develop structured counter-demonstrations collectively in order to respond to Elkan's de-monstration warrants a closer look. In fact, while we have already seen how certain researchers had acquired competence as spokespersons over time, we can now see that these same participants also had the specific skills needed to respond to the detractors of fuzzy logic. The often redundant criticisms with which they had been confronted in the course of their careers had allowed them to identify certain typical de-monstrations and to refine more effective and more readily usable counter-de-monstrative repertories than those resulting from on-the-spot improvisation.

To spare the reader a somewhat laborious description, I shall not go into detail regarding each and every one of the representatives of fuzzy logic we have encountered earlier. It will suffice to look at the cases of David Belmont and Thomas Berger, given the major role these two scholars played in the controversy.

THE FORMATION AND USE OF SEDIMENTED REPERTORIES OF DE-MONSTRATION

David Belmont and Thomas Berger had been serving as spokespersons for fuzzy logic for a long time; as we know, they had been copresidents of the program committee for the FUZCOL conference. Their intervention on the electronic forum had acquired a certain visibility, and they constituted a privileged reference for researchers working in their field. Moreover, the two authors had gotten involved in the defense of fuzzy logic by publishing responses to Elkan's article in a number of journals.

These latter actions simply extended a long series of endeavors that had brought David Belmont and Thomas Berger recognition as representatives of fuzzy logic. Beyond their contributions to the organization of a

number of conferences and colloquia on this theme,[19] they had participated in the creation of an association whose aim was to interest French
industrialists in fuzzy logic. They also coedited a liaison newsletter for
researchers in fuzzy logic. If they had acquired some renown among the
actors in fuzzy logic, this was in part because they had given numerous
papers and published a considerable number of articles on the subject
themselves. They had also published a work of synthesis on fuzzy logic
that was used as a textbook in some American universities.

In the context of FUZCOL, Belmont had summarized his own diverse
activities for me. Like Berger, he worked in a French research laboratory.
Both scholars directed doctoral theses; part of their work was financed
by contracts with industrialists. They held positions as scientific advisers
for major French industrial enterprises. It was not unusual for them to
grant interviews to journalists, and they were frequently cited in newspaper articles and in popular science magazines.

During FUZCOL, several participants had confided to me that they had
great respect and a certain admiration for Belmont's and Berger's work
and erudition, and that this feeling was widely shared. Indeed, during
the conference, I had noticed on several occasions that researchers and
engineers had sought them out to ask for "advice" and for references to
published texts.

The two Frenchmen had long since acquired skill as mediators. Training in engineering followed by the writing of doctoral dissertations and
then postdoctoral appointments in the United States constituted the first
elements in a trajectory that conferred on each of them the mastery of a
vast range of skills and abilities: they could engage in dialogue with engineers and researchers working in many different fields, contribute to publications and debates in France and in the United States, draft symbolic de-
monstrations, and develop expert systems financed by industry contracts.
They were thus in a position to be spokespersons for the actors in fuzzy
logic, actors who were involved for their part in more limited and quite
diversified fields of activity.

Investing energetically in their role as representatives of fuzzy logic,
Belmont and Berger were quick to respond to criticisms of work in this
field. Their political sense was inevitably honed by the hostile reactions
that they had identified or had had to confront throughout their careers,
reactions they were prepared to attribute to a pronounced antagonism to
fuzzy logic. By political sense I mean in particular an ability to grasp the
probable consequences, for the activity of researchers in the field, of the

[19] Their names appeared on the list of organizing committee members on a number of
announcements for conferences and colloquia on fuzzy logic.

diffusion of critical discourses bearing upon fuzzy logic, a capacity to involve themselves effectively in counter-de-monstrative undertakings, and above all to shine a spotlight on the antagonistic actions they were confronting and denouncing. The attribution of negative claims about fuzzy logic to a pronounced hostility toward that theory clearly has to be grasped in its ostentatious dimension, as a de-monstrative register and a "response" mode in its own right.

During FUZCOL, for example, Belmont evoked the *de-monstrations of strength*[20] that he and Berger, along with some of their colleagues, had had to confront up to that point. These de-monstrations both reflected and extended the prevailing critical discourse about fuzzy logic. Belmont also recalled the hostile reactions he had encountered during his first (and unsuccessful) candidacy for a position in a French research institution.

Identifying and denouncing a ferocious opposition to fuzzy logic, Belmont related various instances of blackmail brought to bear for the attribution or maintenance of research positions, "psychological pressures" that had driven some of his colleagues to the breaking point (for example, he had recently received an eight-page "letter full of insults" from a specialist in control theory). He described the opposition of many French academics to the introduction of fuzzy logic as a subject taught at the university level: campaigns to denigrate the field were undertaken, and, in certain forums, articles and papers devoted to fuzzy logic were rejected. But he also evoked boycotts of colloquia and of thesis committees that brought together researchers in fuzzy logic, public accusations intended to destroy the reputations of the leading figures in fuzzy logic, and actions destined to keep specialists in the field from getting industry contracts.

The fear Belmont expressed during FUZCOL that Kosko's theses served the detractors of fuzzy logic and put the whole field at risk, making it less credible, was thus visibly fed by the experience of numerous de-monstrations of strength. We also understand better why Belmont and Berger expended so much energy on responding to Elkan's de-monstration, by intervening in several forums.

---

[20] I am using the expression *de-monstration of strength* in a quite specific sense that does not involve the standard definitions of "strength" or "demonstration," as a way of characterizing interventions based at once on an enterprise of ostentation, through the use of critical registers, and on actions having negative consequences on the activity of the individuals or groups involved (interruptions in chains of credibility, introduction of major obstacles to the pursuit of a research activity involving its most essential material conditions, and so on.) It should be noted that Latour uses this expression in a perceptibly different sense, to qualify a situation in which an actor takes recourse to a large number of allies in order to establish credibility (inscriptions, machines, various assemblies, and so on). More fundamentally, in Latour, the word "strength" appears as a term without a contrary, and the notions of "demonstration" and "strength" cannot be distinguished. See Latour, *Science in Action*. See also

Clearly, the two Frenchmen had acquired specific competencies in counter-de-monstrative action. Facing criticism over time, they had identified certain "standard" de-monstrations and had refined their own counter-de-monstrative repertories. They had trained themselves in dialectical modes of exchange adapted to interlocutors whose activities were quite varied. During FUZCOL, Belmont had thus been in a position to come up with an oral list of typical criticisms addressed to fuzzy logic, whether in the field of artificial intelligence, logic, probability theory, or control theory. An examination of the criticisms presented in the first two of the fields just mentioned will suffice to illustrate the point.

According to Belmont, for instance, a number of proponents of "classical" artificial intelligence reproached fuzzy logic for failing to develop entirely symbolic methods;[21] he reported that many articles in this field "demonized" the use of numeric methods in fuzzy logic, whereas that logic actually used both numeric and symbolic approaches in combination. Belmont claimed, moreover, that fuzzy rules were often identified, quite wrongly, with *weighted rules*.

Similarly, he evoked a set of "typical" criticisms of fuzzy logic formulated by logicians. He even asserted that he was quite invulnerable on this topic, after spending fifteen years with his colleague Thomas Berger "responding to the same mistakes." He identified a first group of criticisms that sought to show the formally contradictory character of fuzzy logic, while erroneously introducing one or more axioms rejected by that logic. Belmont adopted an algebraic reading of this phenomenon, in conformity with the practice of logic that characterized his own intervention on the electronic forum. He declared that these proofs of contradiction had in common the fact that they introduced an axiom *from Boolean algebra*, such as the principle of the excluded middle, whereas fuzzy logic did not have a Boolean algebraic structure and rejected that type of axiom.[22] According to Belmont, many people "had Boolean algebra in their field of vision" and could not adopt any other perspective. It was to this rigid attitude that he attributed the false discoveries of contradictions in fuzzy logic, of which Elkan's proof was just one example.

In the catalog of "blunders" that he said he and Berger had drawn up on the basis of the logicians' attacks, Belmont distinguished a second group of criticisms that had grown out of a more philosophical approach

---

the heuristic virtues of this concept, as it is defined here, in Chandra Mukerji, *Territorial Ambitions and the Gardens of Versailles* (Cambridge: Cambridge University Press, 1997).

[21] He was referring to symbolic manipulations similar to those put into play in the context of instruction in logic such as we saw earlier.

[22] For more details on the notion of Boolean algebra, which has been introduced here in the context of the analysis of teaching logic, see Jean Kuntzmann, *Fundamental Boolean Algebra*, trans. Scripta Technica Ltd. (London: Blackie, 1967 [1965]).

to logic. These latter consisted in pointing out contradictions in the way researchers in fuzzy logic dealt with data and variables linguistically in using terms such as "vague concepts" and "fuzzy variables."[23] According to Belmont, these criticisms generally amounted to "first-degree blunders" or to "delirious extrapolations" that simply missed the fundamental ideas of fuzzy logic.

A third set of critiques brought together the normative condemnations of a semantic—as opposed to a syntactic—approach[24] to logic.[25] According to Belmont, certain logicians were unable to imagine that the notion of interpretation could have a meaning different from the one they attributed to it, and that a logical system could be defined in other than syntactic terms. They asserted that the distinctive feature of a logical system, and the source of its value, lay precisely in its expression in a syntactic form. They insisted that fuzzy logic did not satisfy this condition. For Belmont, an equally normative and even reactionary approach led certain logicians to "have it in for" nonmonotonic logic: they refused to grant it the status of logic inasmuch as the type of inference involved was not deemed to be logical in nature, since it did not possess the "classical" property of being monotonic. Belmont identified that "attitude" with a "lack of open-mindedness" and of "tolerance," characterizing the academics in this latter group as "terroricians."

Finally, in the criticisms of certain logicians Belmont had identified "typical" de-monstrations of the contradictory character of fuzzy logic and the repertories of discredit that were based on antagonistic normative practices of logic. He had done similar work in other fields in which he was involved, such as artificial intelligence, probability theory, and control theory. Over a period of some fifteen years, then, David Belmont and Thomas Berger had been able to identify the counter-de-monstrative elements that could be used to oppose their adversaries, both orally and in writing. They had been able to test various registers and gradually improve them, to the point of considering themselves "quite invulnerable."[26]

---

[23] See J. A. Goguen, "The Logic of Inexact Concepts," *Synthese* 19 (1969): 325–73.

[24] The notions of semantics and syntax are endowed with variable meanings in philosophy and in logic and have given rise to many debates. These terms are often used, however, with reference to attention paid respectively to the meaning or to the modes of construction of statements made in a given language.

[25] See Susan Haack, "Do We Need 'Fuzzy Logic'?" *International Journal of Man-Machine Studies* 11 (1979): 437–45. See also the response to Haack's article in John Fox, "Towards a Reconciliation of Fuzzy Logic and Standard Logic," *International Journal of Man-Machine Studies* 15 (1981): 213–20.

[26] This competence was of course not unique to these two researchers. During my investigations I found it present in other "experienced" academics. Such was the case most notably for researchers inclined to give "little talks," often prepared in advance, that could make an

Belmont's gradual development of counter-de-monstrative registers did not proceed from a dynamics fundamentally different from the one involved in drafting his de-monstrations. Indeed, he explained that his de-monstrative production also resulted from a long distillation process. He asserted that he generally had to work through several oral presentations, collect and analyze the reactions of his interlocutors, and make important transformations in his texts, before he was "understood" and could publish his work. His de-monstrative and counter-de-monstrative productions alike involved a series of adjustments and revisions of proofs,[27] punctuated by assorted interactions, gatherings of interlocutors, observations of reactions, basic forms of opinion surveys, and textual displays (for example, transparencies presented on an overhead projector).[28]

These data further modify our picture of the way the Elkan controversy unfolded. In the face of Elkan's de-monstration, Belmont and Berger were not obliged to "improvise" a response entirely. Elkan's proof belonged to a class of de-monstrations that they had identified and to which they had been responding for a long time. From the start of the controversy, these two Frenchmen had considerable skill in the matter at hand. While Elkan's publication of criticisms of fuzzy logic was apparently his first, Belmont and Berger for their part had "a long track record."

The participants in the debates were thus endowed with unequal resources from the outset. They were much less interchangeable than it might have seemed at first. A posteriori, the de-monstrative activity of the protagonists in the controversy seems to have taken on quite diverse forms. The various participants were not "playing the same game." The interventions of Belmont and Berger in the debates were associated with the display of powerful stakes, with the description of an experience of *de-monstrations of strength* and of an anticipation of the risks that Elkan's de-monstration posed for the pursuit of research in fuzzy logic. Their de-monstrative production stemmed from a less isolated and less contingent undertaking than that of some of their interlocutors. Moreover, the emergence of Elkan's proof itself appears to be a less exceptional event than might have been supposed, inasmuch as it was not without precedents.

---

impression. To allow their authors to *shine* in public, these brief talks generally included jokes and other anecdotes.

[27] It is useful, however, to distinguish between the notions of de-monstration and counter-de-monstration, if only to be able to distinguish the de-monstrations that are presented essentially as "responses" or "counterthrusts" from those that do not stem from such a thematics.

[28] Once again we can assess the pertinence of the descriptions offered by De Millo, Lipton, and Perlis, and of the developments that can be added to these descriptions. The three scientists in fact evoked the essential role played in the dynamics of writing and rewriting by the interactions in which the authors of proofs were engaged. See De Millo, Lipton, and Perlis, "Social Processes."

In the context of the "discussions," the formulation of a de-monstration constituted an act corresponding to variable practices. Intervening in the debates consisted, for some, in drawing on highly sedimented demonstrative repertories, in the context of a long-term commitment to the activity of "defending" fuzzy logic. For others, this undertaking represented a more contingent and improvised action. In this instance, even if the de-monstrations appeared a priori to be isolated cases, they had to be apprehended in the context of the long-term dynamics that led researchers endowed with comparable skills to formulate concordant criticisms.

In the face of the sedimented repertories of participants like Belmont and Berger, we may well ask what de-monstrative approaches Elkan put to work after his article was published. We have already seen that he displayed great mobility and a powerful adaptive capacity, especially in his responses on the electronic forum and in his exchanges with Carpenter. He proved to be an ingenious mediator, capable of utilizing a large number of de-monstrative registers and of producing an evolving series of responses adapted either to individuals or to mass audiences.

In response to the sedimented repertories of certain major figures in fuzzy logic like Belmont and Berger, and to the relatively united front with which they countered his de-monstration, Elkan developed differentiated reformulations of his own viewpoints. He appears to have engaged in a significant number of interactions that offered him the opportunity to draw on new supports, or at least to limit the criticisms and even the actions directed against him to some extent. Because he lacked access to many of the resources available to the leading figures in fuzzy logic, who were apparently more familiar with this type of exercise, Elkan drew above all on his talents as mediator to counter his challengers.

ADVANCING ADAPTIVE, POLYSEMIC, AND DIFFERENTIATED DE-MONSTRATIONS

After his proof was published, Elkan seems to have spared no effort to engage in personalized dialogue with certain of his potential detractors and supporters. Electronic messages sent to Carpenter or to me showed how quickly he could reformulate his de-monstration and his theses in order to find "explanations" adapted to his various interlocutors.

Thus, in a message he sent me in February 1994, he said that he was aware of the fact that in one passage of his proof his use of the equivalence between the statements (not (A and not B)) and (B or (not A and not B)) led to the implicit introduction of the principle of the excluded middle. (Let us recall that the implicit mobilization of this principle had been pointed out on the electronic forum.) Elkan was not acknowledging that he had made any sort of "mistake," however. He was primarily intent on specifying the *meaning* he attributed to his de-monstration, at least in the context of our specific interaction and at this particular moment. His undertaking thus was an adapted defense of the text of his original proof.

This phenomenon may come into clearer focus if we take into account the assertion that immediately followed the previous statement. As Elkan put it, his thesis held that an average user of fuzzy logic might implicitly introduce the principle of the excluded middle in this way without realizing it.

By this new formulation, and by bringing a specific "public" into the foreground, Elkan was offering a transposed and personalized image of his de-monstration. The object of his proof, as he presented it to me in this message, now consisted in showing "average users" that introducing a statement equivalent to the principle of the excluded middle into fuzzy logic led to inconsistencies. Such a displacement was possible as soon as he defined an "average user" of fuzzy logic and pronounced judgment on that user's competence: the latter was unaware that such a result would ensue. In this context, Elkan was primarily delivering a pedagogical message.

We were thus far removed from the formulations of our 1992 interview. Even if the two sets of declarations might be entirely compatible, depending on the meaning attributed to them, fuzzy logic had been presented on the earlier occasion as a bearer of inconsistencies; moreover, an early draft of Elkan's proof led to the same result. This only reconfirms the role played in the evolution of his formulations by the debates and interactions in which the author was involved.

However, we can also appreciate the differentiated character of Elkan's reformulations here. It is probable that the turning point observed in his modes of exposition was not reached before the AAAI'93 conference was held. Such in any case is what a *witness* who reported that he was present at Elkan's talk seemed to be saying in a message addressed to the electronic forum: "In the Q&A after Elkan's talk at AAAI, I suggested he couldn't have [the] excluded middle. He saw no reason why he couldn't."[29]

Other elements underline Elkan's capacity to reformulate the stakes of the debates and of his preceding interventions in response to the criticisms with which he was confronted. For example, in replying to an electronic message from me in early 1994, Elkan expressed agreement with the idea that his proof was elementary from a mathematical standpoint, but he maintained that it was not elementary from the standpoint of the social elaboration of knowledge, as was demonstrated by the debate to which

[29] Bob Dalton, August 31, 1993, cited in message 1823.

it had given rise. Moreover, he proposed to contact the editor of a forth-coming special issue of the journal *IEEE Expert* that was to be devoted to the debate, with the prospect of my participating in it myself, if I was prepared to bring to bear my own viewpoint as a sociologist of science.

In Elkan's eyes, what was at stake in his de-monstration could hence-forth be placed in a visible way on the terrain of the "social elaboration of knowledge." The diversity of viewpoints expressed on the electronic forum had in fact brought to light the extent to which the grasp of the presuppositions and outcomes of his proof and the definition of fuzzy logic were not "self-evident." Elkan was thus ready, at least in the specific context of our interaction, to locate the value of his de-monstration not so much in a potentially astute development of his reasoning or in the strength of his result as in the sociohistorical significance he attributed to it. My own participation in the special issue of the journal opened up the prospect of a reinforcement of this thesis.

Here again, we encounter Elkan's ability to evolve in his interlocutor's area of expertise by combining a large number of de-monstrative registers and skills. His interest in the sociology of science allowed him to see how an analysis coming out of this research field could help underline the value of his de-monstration, in the face of the criticisms leveled by several de-fenders of fuzzy logic. We can appreciate anew Elkan's gifts as a mediator, employed here to respond to his challengers.

This aptitude was a nontrivial resource. Obtaining *public* expressions of support could be of unmistakable value for Elkan after the appearance of his article. Until the publication of the special issue of *IEEE Expert*, Elkan seemed relatively isolated in the forums in which he was participat-ing or scheduled to participate. In particular, I was able to find no trace of any supportive reaction on the part of the AAAI'93 organizers while the controversy was unfolding. Furthermore, messages in defense of El-kan's text were not exactly legion on comp.ai.fuzzy.

We must also note that, shortly before the AAAI conference was held in July 1993, Elkan sent me an e-mail whose tenor warrants close atten-tion. He told me that his draft article on fuzzy logic was now ready to be published in the conference proceedings. He added that the conference was going to offer a veritable "firestorm" of critical articles on fuzzy logic.

As it happens, we know that Elkan published the only paper (beyond a brief presentation of the way a fuzzy controller worked) on the theme of fuzzy logic in the AAAI'93 proceedings. After the fact, in an e-mail he sent me in early 1994, Elkan commented on this unexpected result, expressing regret that no other article on fuzzy logic had been presented at the conference. He noted that his own article had on the whole given rise to many hostile reactions, but that it had also been greeted favorably by people who worked on expert systems.

Thus the anticipated "firestorm" of critical articles on fuzzy logic had not occurred. It is probable that Elkan was referring, before the AAAI'93 meeting, to projects that did not lead to concrete drafts or accepted papers. To be sure, the absence of publication of such texts is hardly surprising, given the "genre" of communications presented at this type of conference. The presentation of critical notes was a rare occurrence.

Did this phenomenon stem in part from the fact that the presentation of current research work was easier to justify and less "risky" than that of a critical essay? Did it originate at least in part from the difficulty of transforming jokes made around the water cooler into a set of elaborate written critiques that could withstand the test of critical readings by the authors in question and third parties? However this may be, contrary to what he might have expected, Elkan found himself relatively alone facing criticism by a certain number of researchers in fuzzy logic.

Given the obvious lack of public support from his colleagues, for which a simple "favorable reception" (perhaps involving indications of approval expressed solely in private) was substituted, it might have appeared quite valuable to Elkan to benefit from new support and to have to confront only limited hostile reactions. The fact that he evinced great mobility in the *accompaniment* of his article seems to have helped produce such an outcome. Let us recall that the term "accompaniment" serves to characterize an undertaking that consisted, for Elkan, in becoming the spokesperson for his own article on numerous occasions: by reformulating his theses orally and in writing, by producing *new* de-monstrations, and by engaging in a variety of interactions in order to produce points of *view* that took shape in relation to the evolving exchanges.

In the context just evoked, the points of agreement Elkan found between himself and an important figure in fuzzy logic whom I shall call Peter Turner must have been quite valuable for him. In an e-mail he sent me in March 1994, Elkan affirmed that he had received a critical but friendly letter from Turner. A few weeks later, he reported, they had had the opportunity to talk for several hours during a conference in which they had both participated. Elkan declared that Turner had lived up to his reputation, that he was a truly kind man and an authentic gentleman. Elkan indicated that in a rather subtle way Turner had reached a point of view fairly close to his own and was now in agreement with many of Elkan's theses. Elkan added that, after all, he had always asserted that fuzzy logic was a valuable technology. He noted, finally, that he and Turner would perhaps write an article together in the coming months.

On the basis of the available data, it appears that Turner had been absent from the debates up to this point; he was certainly absent from the electronic forum. Rather than expressing his criticisms in a public way, this academic thus seems to have opted at least initially for private exchanges with Elkan, through the letter he sent Elkan and through the dialogue they had begun at the conference. A face-to-face encounter between two first-rate mediators had thus clearly come about.

Had Turner attempted in this way to restrain Elkan's critical ardor by inflecting his positions? Had he tried to neutralize him in part, or even to find a new resource with which to counter the critics of fuzzy logic? Whatever the case, even as he found points of agreement with Turner, Elkan for his part no doubt complicated his own formulations in the same way he would complicate the work of his detractors, if the result of this exchange were made public. The prospect of publishing an article with Turner—a potential de-monstration of legitimacy for Elkan vis-à-vis the defenders of fuzzy logic, narrowing a priori the latter's room for maneuver in the formulation of their critiques—could well have represented a positive step from Elkan's standpoint.

The adoption of certain modes of *behavior* seems to have played an important role in the "success" of such a transaction. The institution of a convivial atmosphere undoubtedly allowed the participants to bracket their differences in order to advance in a nuanced way toward an ingenious compromise. The success of this arrangement owes perhaps as much to the finesse manifested by each of the two interlocutors as it does to their more specific aptitudes for temporarily attributing full importance to the conditions of the exchange. An understanding about how to carry on a dialogue and what constitutes ethical behavior might be provisionally substituted for a disagreement on the points that separated them. This may well be why Elkan stressed his interlocutor's "human qualities." For in the end such attributes may correspond to attitudes that supported the actors' efforts, however laborious, to reach common positions in their debates.

A Goffmanian approach to modes of interaction may be useful here again as a way of grasping the unfolding of the controversy,[30] even if it is only partly satisfying. In fact, as we have seen, we are not simply looking at interactions among individuals here. The individuals in question were often confronted initially with written texts, and as a general rule these texts were an integral part of the interactions, simultaneously constraining and allowing them. We could not have known *on what* the interactions bore nor *in what* they consisted, if we had not also included these texts among our objects.

However, we must note that the opportunity Elkan had, after the publication of his proof, to come into contact with an actor in fuzzy logic such

---

[30] Let us note here, too, the correlation between the attribution of moral qualities to individuals and the constitution of an agreement about scientific statements. For a broader treatment of this question, see Shapin, *Social History of Truth*.

as Peter Turner was certainly not an isolated one. It would seem that
he had multiple possibilities to interact and produce personalized and
adaptive reformulations of his de-monstration and thereby reach points
of agreement with his various detractors. In early 1994, Elkan asserted
in an e-mail that he had received a number of letters and electronic messages, although none very different from those that had been addressed
to comp.ai.fuzzy.

The letters sent to Elkan provided him with opportunities to engage in
personalized dialogues with his challengers, to adjust his formulations on
a case-by-case basis while specifying the meanings to be attributed to his
de-monstration on each occasion. A new image of Elkan's de-monstrative
approach ultimately emerges from the foregoing descriptions. The author's de-monstrative activity seems to have been powerfully determined
by his significant qualities as a mediator, qualities he used to reach points
of (at least) local agreement. The defense of his original de-monstration
appears strongly marked by adaptive, differentiated, and polysemic representations (both oral and written). Elkan clearly had both the ability
to use a wide variety of modes of accompaniment and considerable mobility in his position-taking.

Thus we cannot settle for a model of rationality celebrating the univocity of the text of a symbolic de-monstration. The proofs taken as our
object were endowed by the actors with multiple meanings, in the context
of adaptive re-presentations.[31] If we had been looking a priori for a single
meaning for Elkan's original de-monstration, taking into account the entire set of his writings and testimony, our quest would have failed. On the
contrary, we had to try to grasp the evolution and the diversity of the representations Elkan gave of his own de-monstrative undertaking in the
course of his multiple interactions. In other words, to grasp such a dynamics, we cannot do without a Goffmanian hypothesis on the actors' lack
of unity in their self-presentation and their presentation of their own
work.[32] Each of Elkan's new texts or declarations incited his audience to
read his original de-monstration in a very particular way. On each occasion, new reading instructions were given that partially determined the
readers' interactions with Elkan's current texts and its predecessors.

It is now time to ask whether we can find traces of such mobility in the
successive versions of Elkan's article and, if so, whether we can evaluate
its impact. We need to find out whether the foregoing analyses can be
confirmed and completed by bringing to light and studying in detail the
re-presentations written down on paper. Above all, we shall have to exam-

---

[31] Once again, see the citation from Proust at the beginning of part 3.

[32] See Goffman, *Presentation of Self*. On this topic, see also the analyses developed in
Bernard Lahire, *L'homme pluriel: Les ressorts de l'action* (Paris: Nathan, 1998).

ine the way the author modifies his formulations from one version to the next, in the face of criticisms expressed by researchers in fuzzy logic on the electronic forum and elsewhere.

Besides, we may wonder whether the controversy was eventually stabilized. I shall attempt to show that it was; I shall try to describe the stabilizing operators at work, and above all to analyze the role played by the various versions of Elkan's article and other texts published in journals as the debates unfolded.

Finally, it will be important to determine whether any one particular version of Elkan's theorem finally succeeded in achieving the status of certified knowledge, and at the same time what precise significance might be attributed to such a notion. These questions require close analysis of the various versions of Elkan's text (and of texts published by his challengers); this will be the focus of the following chapter.

Chapter 7

THE EMERGENCE OF A QUASI-OBJECT AND A COLLECTIVE STATEMENT

RECOURSE TO TACIT MANIPULATIONS: DE-MONSTRATION AS A QUASI-OBJECT

N OW THAT WE have looked briefly at the theses Elkan defended in his AAAI'93 paper and examined the debates to which it gave rise, we can analyze the content of the original article in detail, beginning with the proof of the theorem. This step, an essential one if we are to be able to circumscribe the role played by the author's text in the developing controversy, will also make it easier to see just how successive versions of Elkan's article—and especially the symbolic de-monstrations—were transformed.

First, let us recall how the proof was introduced. In a paragraph titled "A Paradox in Fuzzy Logic," placed immediately after the general introduction,[1] Elkan announced that he was going to demonstrate that a "standard" version of fuzzy logic, described as a formal system endowed with a specific axiomatics, consisted in fact of a logical system with two truth values, whereas it was usually presented—always, according to the author—as a logical system authorizing an indefinite variety of truth values. Four axioms constituted the point of departure for the demonstration (here, "$\wedge$" represents the logical connector "and," "$\vee$" the connector "or," "$\neg$" the connector "not," and "$t(\ )$" represents the truth value of the assertion in parentheses):

$$t(A \wedge B) = \min \{t(A), t(B)\}$$
$$t(A \vee B) = \max \{t(A), t(B)\}$$
$$t(\neg A) = 1 - t(A)$$
$$t(A) = t(B) \text{ if } A \text{ and } B \text{ are logically equivalent.}$$

Elkan then stated the following theorem: "For any two assertions A and B, either $t(B) = t(A)$ or $t(B) = 1 - t(A)$." Elkan "immediately" deduced

---

[1] Elkan, "Paradoxical Success," August 1993, p. 698.

from this that fuzzy logic, as a formal system characterized by the four preceding axioms, was "in fact" a logical system with two truth values (zero and one).[2]

Before de-monstrating this result, Elkan indicated that he was defining logical equivalence "according to the rules of classical two-valued propositional calculus." He added that the use of alternative definitions of logical equivalence would be considered later in the article. Thus we can localize a first passage that was ultimately subject to controversy.

At this point, Elkan was ready to provide the proof of his result. Here is its content (the numbers in square brackets are my own annotations):[3]

Let A and B be arbitrary assertions.
Consider the two sentences $\neg(A \wedge \neg B)$ and $B \vee (\neg A \wedge \neg B)$.
These are logically equivalent, [1]
so $t(\neg(A \vee \neg B)) = t(B \vee (\neg A \wedge \neg B))$. [2]
Now $t(\neg(A \wedge \neg B)) = 1 - \min \{t(A), 1 - t(B)\}$ [3]
   $= 1 + \max \{-t(A), -1 + t(B)\}$ [4]
   $= \max\{1 - t(A), t(B)\}$ [5]
and $t(B \vee (\neg A \wedge \neg B)) = \max \{t(B), \min \{1 - t(A), 1 - t(B)\}\}$ [6]
The numerical expressions above are different if
$t(B) < 1 - t(B) < 1 - t(A)$ [7]
that is if $t(B) < 1 - t(B)$ and $t(A) < t(B)$, [8]
which happens if $t(A) < t(B) < 0.5$. [9]
So it cannot be true that $t(A) < t(B) < 0.5$. [10]

Now note that the sentences $\neg(A \wedge \neg B)$ and $B \vee (\neg A \wedge \neg B)$ are both re-expressions of the material implication $A \Rightarrow B$. One by one, consider the seven other material implication sentences involving A and B:

$\neg A \Rightarrow B$
$A \Rightarrow \neg B$
$\neg A \Rightarrow \neg B$
$B \Rightarrow A$
$\neg B \Rightarrow A$
$B \Rightarrow \neg A$
$\neg B \Rightarrow \neg A$

By the same reasoning as before, none of the following can be true: [11]

$1 - t(A) < t(B) < 0.5$
$t(A) < 1 - t(B) < 0.5$
$1 - t(A) < 1 - t(B) < 0.5$
$t(B) < t(A) < 0.5$

[2] Ibid.
[3] Ibid., pp. 698–99.

$$1 - t(B) < t(A) < 0.5$$
$$t(B) < 1 - t(A) < 0.5$$
$$1 - t(B) < 1 - t(A) < 0.5$$

Now let $x = \min\{t(A), 1 - t(A)\}$ and let $y = \min \{t(B), 1 - t(B)\}$.

Clearly $x \leq 0.5$ and $y \leq 0.5$ [12]
so if not $x = y$, then one of the eight inequalities
  derived must be satisfied. [13]
Thus $t(B) = t(A)$ or $t(B) = 1 - t(A)$. [14]

Let us note first of all that the point of departure for the proof is a logical equivalence (see [1]) that, as we have seen, gave rise to numerous debates on the electronic forum. In this first step in the de-monstration, the author chooses simply to show two statements as equivalent. He *exhibits* an equivalence without explaining why he attributes this status to the pair of statements in question. The structure of the passage illustrates quite well the extent to which the de-monstration is composed of a series of *monstrations*, that is, moments in which, implicitly, the author considers it sufficient to *show*, without providing justifications for what he *shows as such*. Challenging a step in a de-monstration, as we have seen on comp .ai.fuzzy, consists precisely in demanding justifications, or in producing a textual fragment intended to refute what is presented as "self-evident."

What follows in the de-monstration brings other exhibitions into play. It appears first of all that Elkan implicitly uses the fourth axiom to *translate* the preceding logical equivalence into a numeric equality, the equality between two identical truth values (see [2]). In so doing, the author is introducing a *calculus*. An examination of this passage will clarify the sense in which we are dealing with a presentation that can be characterized as formalist: the author *makes visible* the legitimate character of the translation that is carried out in this step of the de-monstration. This operation is inscribed a few lines earlier, in the statement of the fourth axiom, very well exposed to the reader's view (and also to the author's, for the purpose of self-verification).

This fact certainly explains in part why Elkan does not refer explicitly to the use of the fourth axiom. The author presumably supposes that the reader is *competent* to make observations and identifications of this sort. An appeal to a form of trust on the reader's part doubtless comes into play here as well. Subjected to temporal constraints themselves, readers may not carry out such a task systematically; they may rely on the author instead.[4] Conversely, mentioning the recourse to the fourth axiom could have harmed the presentation's "elegance" and consequently its author's

---

[4] See Davis and Hersh, "Rhetoric and Mathematics."

reputation and even his distinction.[5] It might also have compromised the economy of the article, which was subject to editorial constraints, notably a limitation on its length.

The possibility of relying on implicit manipulations and on the reader's competence may also be seen as a consequence of the adoption of priorities in the way logical arguments are written. The writer of a proof makes more or less improvised, more or less habitual and sedimented choices in the drafting process, choices that institute or help stabilize hierarchies among logical practices endowed with a more or less "universal" status. Here as elsewhere, we need to look at the various ways logic is written and read, whether through teaching, local practices, or practices that are more or less extended throughout networks, in order to seek the emergence of a *borderline* between operations deemed to require development and explicit references on the one hand, and textual fragments deemed self-sufficient that call potentially for tacit manipulations on the other. To be sure, for a thorough understanding of the reasons why Elkan chose to rely on tacit manipulations in the de-monstrative step we are considering, we would have to be more familiar with his own trajectory and we would need to undertake an extremely fine-grained analysis that would take us well beyond the framework of the present study. For our purposes, it is most important to stress Elkan's recourse to tacit manipulations in his de-monstrative practice.

Such a prism makes it possible to account for the next step in the demonstration (see [3]). At this point, in order to produce a new statement, the author seems to use the first and third axioms. To obtain equations [4] and [5], we may say that he tacitly uses wholly "standard" calculation rules[6]—which he himself helps to stabilize by using them. He then appears to be implicitly drawing on the first three axioms in order to obtain equation [6]. Steps [7], [8], and [9] seem to be obtained through recourse to "standard" calculation rules analogous to the preceding ones, and [10] by capitalizing on the transformations of the first numerical equation (located in [2]).

With step [11], the author asserts that the application to seven other logical equivalences of a reasoning process "similar" to the preceding one shows that seven new double numerical inequalities are not verified. The appeal to the reader's trust, whether it is motivated by the weight of editorial imperatives, the anticipation of temporal constraints on the reader's part, or the definition of priorities in the drafting of the proofs, is once again flagrant.

---

[5] On the search for distinction in the practice of mathematics, see Rosental, "Art of Monstration," and Davis and Hersh, "Rhetoric and Mathematics."

[6] To use more canonical terminology, what we have here are rules for calculation with real numbers.

Using the "standard" rules for calculation without mentioning them, and imposing limits on the range of truth values for a proposition without justifying this step, Elkan then arrives at two inequalities (step [12]). Still seeming to be relying tacitly on "standard" calculation rules, he then formulates a statement in prose (step [13]). Apparently using a form of reductio ad absurdum, he finally concludes with the previously announced result.

Thus we see how Elkan's proof seems implicitly to mobilize various manipulations at every step in the de-monstration. I have attempted to make these manipulations explicit, at least in part, as a participant observer[7] endowed with my own competence as reader. However, I do not claim to be offering an objectivizing viewpoint on the tacit manipulations required ("in the absolute") in order to make the de-monstration "hold together" ("ideally"). I simply want to make clear what a reader such as myself can "see" as implicit manipulations that are used to move from one step of the de-monstration to another.

This point must be emphasized, for while the participants in the electronic forum marked their disagreement by stressing Elkan's recourse to various implicit manipulations in his proof, it is probable that they "saw" other tacit operations in the proof that remained unchallenged. Such was certainly the case, for example, for certain operations obeying "standard" calculation rules. In other words, as they read through Elkan's proof the participants in the controversy undoubtedly experienced a greater number of "visions"—glimpses of these other tacit operations—than were brought into play in the debates.

Still, the quite particular character of the reading (a situated reading, if only in historical terms) of this de-monstration that I am proposing needs to be stressed. I do not claim to offer the viewpoint of a (unique) "ideal" reader who is exclusively determined, in a very precise way, by the text itself. Various forms of visual appropriation of Elkan's de-monstration are possible, as we have seen in the previous debates. A semiotic approach to this text needs to take them into account, if it is to avoid crediting signs with mechanical effects that would only reflect the analyst's own reading practices.[8]

In particular, it is important to stress that the borderline between the implicit and explicit domains marked out by my reading is quite specific. What my own competence leads me to characterize, in the passage from one step of the de-monstration to the next, as recourse to implicit manip-

---

[7] My status was that of an observer evolving in the extension of the play of the actors. On this methodological point, see Latour and Woolgar, *La vie de laboratoire*.

[8] This is certainly one of the major difficulties encountered by semiotic studies dealing with mathematics. The problem remains the same when the analyst's descriptions have an ideal aim and do not pretend to be realistic. See again chapter 1, note 15.

ulations, might be considered by others as stemming from explicit operations. Again, the radical divergences in the ways de-monstrations can be seen, divergences that were expressed on comp.ai.fuzzy, attest to this quite clearly.

The position of participant observer that I have just adopted has at least one advantage: it shows that Elkan's de-monstration might also be subject to partially consensual readings, even if the points of agreement are not evoked in the debates. Moreover, this stance has allowed me to problematize the very notion of "implicit manipulation." Let us recall that several elements have been evoked to account for the existence of such manipulations, in particular the following: constraints imposed on the length of the presentation, appeals to a form of trust on the reader's part, a form of distinction in the nonexplicitation of "trivial" manipulations, recourse to the reader's competence to carry out operations that are presumed to be known and accepted by this reader, definitions of priorities in the drafting of proofs, connected with modes of learning logic and with professional practices that may be local or more or less widely extended in networks.

Such phenomena clearly run counter to the spirit of a formalism that seeks to make everything visible. In that ideal framework, a demonstration would consist in a series of *monstrations*, of moments (or privileged passages) in which the author is content to show without having to justify what he is *showing as such*.

The foregoing analyses provide a certain number of elements that make it easier to understand what underlies the "necessity" for the various moments of Elkan's proof from the actors' point of view. It must be emphasized that efforts to solve this sort of problem often suffer from a lack of empirical data or from a *stylization* of the examples presented. Very polarized analyses may result from these approaches: thus we find both highly objectivizing[9] and highly sociologizing explanations.[10]

This case study shows the usefulness of a *nonfrontal* approach to the question. We see that the various stages of Elkan's proof appear necessary to some readers but not to others, especially in relation to the reading practices to which they are subjected. This issue benefits in fact from being repersonalized (what is or is not necessary for which actor or which group of actors?). It must also be remediatized: by what media, within what

---

[9] See especially Husserl, *Logical Investigations*.

[10] Again, see Davis and Hersh, "Rhetoric and Mathematics." One of the most famous treatments that attempts to escape from this polarization, but that unfortunately stumbles in its de-monstrative efforts over the stylization of the cases examined, certainly lies in Wittgenstein's work. On this subject, see Bloor, "Wittgenstein and Mannheim." See also the attempt to escape from such polarization in Pickering and Stephanides, "Constructing Quaternions."

testing mechanism do the steps in the de-monstration strike readers as necessary or not? What consensuses and what competences are mobilized?

For example, the study of why the step in Elkan's de-monstration corresponding to the passage from statement [1] to statement [2] appears necessary to certain readers can be pushed quite far. We are dealing with a transformation objectively authorized by a system of axiomatic *signalization*, incomplete instructions in the reading of the proof. Still, the recognition of this translation presupposes a global adherence to this type of systemic approach. As we have seen, it implies various forms of competence and complicity on the reader's part. But it also refers to the acceptance of the editorial constraints to which the author had been subjected, to the adoption of a specific formalist approach, and so on.

These data confer on Elkan's de-monstration all the properties of a quasi-object: an entity susceptible of being split into an object pole and a social pole.[11] This reality came across clearly in the debates on the electronic forum. Certain participants associated Elkan's de-monstration with an object pole, by declaring it objectively correct. Others placed it at a social pole, seeing in it essentially the mark of one community's hostility to fuzzy logic. In the course of the exchanges, the combinations of *visions* associated with the proof led to a constant displacement of its representation along an axis drawn between these two poles. We see how the search for what makes one or another step in the proof necessary for the actors leads to a perhaps unanticipated study: tracing the trajectories of a quasi-object.

Having examined the text of the proof Elkan published in the AAAI'93 proceedings, we are now in a position to analyze the transformations made in its successive versions. Let us begin by comparing this de-monstration with the one intended for publication in the journal *IEEE Expert*.

DEFENDING A PROOF BY REFORMULATING IT

The text Elkan intended to publish in *IEEE Expert* was available on the electronic network starting in November 1993.[12] Let us recall that American academics and computer scientists were generally familiar with the

---

[11] See Bruno Latour, "One More Turn after the Social Turn: Easing Science Studies into the Non-modern World," in *The Social Dimensions of Science*, ed. Ernan McMullin (Notre Dame, Ind.: Notre Dame University Press, 1992), pp. 272–94. See the introduction of the notion in Michel Serres, *Statues* (Paris: François Bourin, 1987). This notion of a quasi-object must not be confused with the one introduced by Carnap, who gives it a different meaning. On this subject, see Boll and Reinhart, *Histoire de la logique*.

[12] Charles Elkan, "The Paradoxical Success of Fuzzy Logic," *IEEE Expert*, August 1994, pp. 3–8.

procedures for downloading data on this network. In addition, the author and several participants made many references to the November 1993 version on comp.ai.fuzzy. Endowed with a certain visibility, this text constituted an important element in Elkan's de-monstrative efforts to stand up to his challengers until early 1994.

Nuances and Precautions

The summary of the first version of Elkan's article reads as follows:

> This paper investigates the question of which aspects of fuzzy logic are essential to its practical usefulness. We show that as a formal system, a standard version of fuzzy logic collapses mathematically to two-valued logic, while empirically, fuzzy logic is not adequate for reasoning about uncertain evidence in expert systems. Nevertheless, applications of fuzzy logic in heuristic control have been highly successful. We argue that the inconsistencies of fuzzy logic have not been harmful in practice because current fuzzy controllers are far simpler than other knowledge-based systems. In the future, the technical limitations of fuzzy logic can be expected to become important in practice, and work on fuzzy controllers will also encounter several problems of scale already known for other knowledge-based systems.[13]

Here is the second version of the summary:

> Applications of fuzzy logic in heuristic control have been highly successful, but which aspects of fuzzy logic are essential to its practical usefulness? This paper shows that an apparently reasonable version of fuzzy logic collapses mathematically to two-valued logic. Moreover, there are few if any published reports of expert systems in real-world use that reason about uncertainty using fuzzy logic. It appears that the limitations of fuzzy logic have not been detrimental in control applications because current fuzzy controllers are far simpler than other knowledge-based systems. In the future, the technical limitations of fuzzy logic can be expected to become important in practice, and work on fuzzy controllers will also encounter several problems of scale already known for other knowledge-based systems.[14]

In comparing these two versions, we cannot help noting the euphemization of the criticisms of fuzzy logic leveled by the author after the start of the controversy. In the November 1993 version, Elkan began by emphasizing the "successes" of fuzzy logic ("applications of fuzzy logic in heuris-

---

[13] Elkan, "Paradoxical Success," August 1993, p. 698.
[14] The version dated November 1993. This version was accessible from the site cs.ucsd.edu under the heading pub/paradoxicalsuccess.ps via an FTP command. From here on, to distinguish this version from other versions, I shall systematically give the month as well as the year in each case.

tic control have been highly successful"), whereas in the August 1993 version this affirmation is buried in the middle of the text.

Aware as we are of the debates on comp.ai.fuzzy, we are in a position to note that Elkan's most "devastating" criticism, the one tied the most directly to the statement of his theorem, has been highly attenuated: "a standard version of fuzzy logic collapses mathematically to two-valued logic" thus becomes "an apparently reasonable version of fuzzy logic collapses mathematically to two-valued logic."

In one of Elkan's messages to me, the author described the equivalence noted in [1] in the following terms: it was one that an "average user" of fuzzy logic might introduce without realizing that he was thereby mobilizing the principle of the excluded middle, a source of inconsistencies. Consequently, his de-monstration could be said to have the following significance: it seemed to show "in fact" that the introduction of equivalences of "usual" or "reasonable" use led to inconsistencies in fuzzy logic. This is indeed one way to read the allusive expression "apparently reasonable," used by Elkan in November 1993 in the context of a polysemic demonstration.

Given the vagueness of its expression, the general thrust of Elkan's demonstration was maintained: it was still a matter of contrasting "the limits" of the formal system of fuzzy logic with the success of its "applications." But the new formulations offered Elkan's various readers distinct messages capable of "satisfying" them better. A researcher in fuzzy logic who was following the debates could find a more "modest" and more "accurate" result. In contrast, a reader "in a hurry" and critical of fuzzy logic could find a reaffirmation of the correctness of Elkan's theorem, without necessarily grasping the nuances added from one version to the next. Bringing to light the potential effects of the ambiguity of a simple expression enables us to begin to see how polysemic reformulations could generate points of agreement both with Elkan's previous challengers and with his allies, and could help stabilize the controversy.

Let us now examine the other alterations made in the summary. The assertion that fuzzy logic does not constitute a formalization adequate for reasoning under uncertainty, in the framework of expert systems, is abandoned in favor of a more cautious declaration. The author now stresses the lack of publications of *reports* on this practice—namely, the use of fuzzy logic in expert systems to carry out reasoning under uncertainty.

In addition, instead of stressing inconsistencies proper to fuzzy logic, in the November 1993 version Elkan simply evokes the existence of limits proper to this formalism. "We argue that the inconsistencies of fuzzy logic have not been harmful in practice" thus becomes "It appears that the limitations of fuzzy logic have not been detrimental in control applications." Let us note that "We argue," which engages the responsibility of the author, becomes the impersonal and less affirmative. "It appears." The ambiguity of such an expression once again opens up the prospect of delivering distinct messages to different readers.

Only the final sentence of the summary remains unmodified. This result is hardly surprising in that, at this level, Elkan's criticisms were already measured in August 1993: the problems he predicted for complex applications of fuzzy logic were balanced against the ones he foresaw for complex systems based on classical artificial intelligence.

In the end, comparing the two versions of this summary allows us to measure the effects of the exchanges that took place between July and October 1993 on the electronic forum. The defense of Elkan's demonstrations visibly passes through more cautious formulations in the new version of the text. The introduction of polysemic turns of phrase appears apt to deliver reassuring messages to the author's detractors while at the same time conveying an image of firmness to his earlier supporters. We thus begin to see how the author has been able to set up a device that can "satisfy" researchers in "classical" artificial intelligence and that can simultaneously minimize or even on occasion "settle" disagreements like the ones between Elkan and his challengers.

The second version of the article thus bears the trace of the flexibility Elkan manifested in other contexts after the publication of his text. The relevance of one of Lakatos's theses[15] to Elkan's case is now more apparent: the "formal" presentation of a result comes only at the end of a process of stabilization and stands as a record of past negotiations.

This said, we must note the existence of perceptible divergences between the formulations in the summaries and those in the texts themselves. This observation is even more striking for the second version of Elkan's article than for the first. Such divergences also help contribute to the defense of the de-monstration. A double-barreled mechanism is at work.

The summary, which constitutes the most visible part of the article, delivers to Elkan's detractors—especially in the second version—a reassuring message based on more cautious formulations than those found in the body of the text. In the text itself, as we shall see, certain expressions offer elements apt to satisfy readers who are critical of fuzzy logic. The polysemic character of some of these expressions limits the possibility of finding contradictions between the summary and the body of the text, even as it allows Elkan's critics to believe that their objections have been taken into account and have led to more nuanced statements.

Let us examine these modifications in some detail. If we look only at the section titles, the two versions of the article seem identical; in

---

[15] Lakatos, *Proofs and Refutations*.

both cases, we find the following: (1) "Introduction"; (2) "A Paradox in Fuzzy Logic" [the section that presents the proof of the theorem]; (3) "Fuzzy Logic in Expert Systems"; (4) "Fuzzy Logic in Heuristic Control"; (5) "Recapitulating Mainstream AI"; (6) "Conclusions." In the text itself, however, important transformations have been made in the second version.

A Polysemic Textual Device to Stabilize Debates

The changes made in Elkan's introduction are comparable to those we saw in the summary.[16] First of all, a somewhat triumphal assertion ("Our conclusions are based on a new mathematical result") is abandoned in favor of a more neutral declaration ("The conclusions here are based on a mathematical result"). This change comes in response—at least in the eyes of those who followed the entire set of debates—to certain criticisms expressed on the electronic forum. The fact that the result already existed in generalized forms had been evoked in that context.

Furthermore, the statement that "section 2 **proves** and discusses the theorem mentioned above, which is that only two truth values are possible inside a **standard** system of fuzzy logic" was replaced in November 1993 by the following: "Section 2 **states** and discusses the theorem mentioned above, which is that only two truth values are possible inside an **apparently reasonable** system of fuzzy logic." The substitution of the term "apparently reasonable" for "standard" may once again suggest to some readers that the limited scope of the theorem has been acknowledged, without any challenge to its "correctness." But to the extent that lesser importance is attributed to the results of the proof, the latter is allotted a more peripheral place. The proof has been removed from the second section and placed in an annex. This explains why the author declares, in the second version, that only the *statement* of the theorem (and not its proof) is given in the second section.

However, we must note that certain assertions not fully in keeping with the new summary are maintained. Indeed, these statements counterbalance the effect produced by the November summary's more measured formulations. They may give the detractors of fuzzy logic the idea that a critical stance is still in force. This can be observed if we examine two sentences excerpted from the introduction that are common to both versions of the article: "Fuzzy logic methods have been used successfully in many real-world applications, but the coherence of the foundations of fuzzy logic remains under attack. . . . The tentative conclusion is that suc-

[16] See Elkan, "Paradoxical Success," November 1993, pp. 1–2.

cessful applications of fuzzy logic are successful because of factors other than the use of fuzzy logic."[17]

A certain firmness of tone is perceptible in this passage, in both versions of the article. The author seems to be denouncing the "incoherence" of fuzzy logic[18] and claiming that fuzzy logic has nothing to do with the success of the applications that have been attributed to it.

The second version of the text thus manifests a highly nuanced construction. From November 1993 on, then, with the help of the new version of his article, Elkan was in a position to underline various significations of "his" de-monstration, depending on which interlocutors he was addressing. To this end the new text offered him considerable freedom to maneuver. To be sure, its polysemy could be used by his detractors to shore up criticisms and pinpoint inconsistencies. But the polysemic character of certain expressions was such that this type of response was ultimately difficult to make.

As for the section of the article devoted to the proof, its title remains unchanged, and the result to which the de-monstration leads is still characterized as a paradox. As we have seen, the most immediate modification of this section lies in the transfer of the proof to the annex. But this is not the most fundamental change. Certain theses are maintained in the new version even though they are now accompanied by a number of stylistic precautions. Readers with the skills to decipher them must have been able to see the presentation of a less ample result than the one that appeared to be defended in the first version of the article. Readers without those skills were less likely to perceive the differences. In fact, the symbolic proof as such was hardly modified. The changes primarily affected the accompanying text, and especially the passages that had been the object of the most discussion on the electronic forum.

Let us look first at the modifications introduced in the preamble. The very statement of the theorem is altered in the second version of the article.[19]

First version: Theorem 1: For any two assertions A and B, either $t(B) = t(A)$ or $t(B) = 1 - t(A)$.

Second version: Theorem 1: **Given the formal system of Definition 1, if** $\neg (A \wedge \neg B)$ **and** $B \vee (\neg A \wedge \neg B)$ **are logically equivalent, then** for any two assertions A and B, either $t(B) = t(A)$ or $t(B) = 1 - t(A)$.

$$(B \vee \neg A) \wedge (B \vee \neg B)$$

[17] Elkan, "Paradoxical Success," August 1993, p. 698.

[18] We may wonder whether this sentence does not bear several meanings. One of them would perhaps be the following: *other* authors would stress the incoherence of fuzzy logic (while the author of the article is not taking a stand on the subject).

[19] See Elkan, "Paradoxical Success," August 1993, p. 698, and November 1993, p. 3.

$$\neg (A \wedge \neg B) \Leftrightarrow \neg A \vee B$$

In the second version, the "apparently reasonable" logical equivalence that earned Elkan so much criticism is no longer introduced in the same way. The author attributes to it a clearly hypothetical character, as attested by the use of the conditional, "if . . . then." By this change in the way the theorem is formulated, the de-monstration responds to one of the principal objections leveled against it and becomes more "resistant." For anyone who reads only this second version, Elkan cannot be accused of having committed an elementary error, that of unwittingly introducing the principle of the excluded middle. For readers who grasp what is at stake in this nuance, he is only de-monstrating a weaker result than the one that might have been envisaged initially: in fuzzy logic, one cannot introduce the logical equivalence in question without ending up with inconsistencies. But this reformulation has still other effects. Given the relatively technical nature of the argument, the theorem should not appear weaker in scope to numerous partisans of classical artificial intelligence.

From this point on in the second version of his article, Elkan's de-monstration takes the form of an openly hypothetical deductive undertaking. A number of markers turn up to accredit the idea that the author is fully aware of the hypothetical character of the equivalence he is introducing, as the transformation of the following passage illustrates:[20]

First version: Proof: Let A and B be arbitrary assertions.
Consider the two sentences $\neg (A \wedge \neg B)$ and $B \vee (\neg A \wedge \neg B)$. These are logically equivalent, so $t(\neg (A \wedge \neg B)) = t(B \vee (\neg A \wedge \neg B))$.

Second version: Proof: Given the assumed equivalence, $t(\neg (A \wedge \neg B)) = t(B \vee (\neg A \wedge \neg B))$.

However, other adjustments can be observed in the prolegomena to the proof. The revisions mainly involve the author's comments on the signification and scope of the result de-monstrated.

Visibly integrating certain criticisms expressed on the electronic forum concerning the axiomatic system he was associating with fuzzy logic, Elkan adopts a certain number of stylistic precautions in the second version of his text. The introduction of this system of axioms, now regrouped under the heading "Definition 1," is the object of the following reservations: "**Depending on how the phrase 'logically equivalent' is understood, Definition 1 yields different formal systems.** A system of fuzzy logic is intended to allow an indefinite variety of numerical truth values. However, **for many notions of logical equivalence** only two different truth values are possible given the postulates of Definition 1."[21]

[20] See Elkan, "Paradoxical Success," August 1993, p. 698, and November 1993, p. 16.
[21] I have indicated the reservations in bold type. See Elkan, "Paradoxical Success," November 1993, pp. 2–3.

In this smooth, allusive formulation, an important "clarification" is introduced, responding in particular to objections expressed on comp.ai .fuzzy concerning the notion of logical equivalence used in the proof. On the basis of this passage, Elkan's critics can no longer fault him for being unaware of the fact that a notion of logical equivalence affects the nature of the formal system that uses it. Still, throughout the evasive affirmation according to which "several" notions of equivalence lead to the same result, the text remains in a position to convince certain readers (especially readers unfamiliar with the debates) that the theorem has broad "generality."

We see here again how the polysemic character of the reformulations allows the new version of Elkan's de-monstration to convey different messages to dissimilar sets of readers. Given the heterogeneity of the readers' qualifications and the differences in their level of investment in the debates, the proof can be accorded multiple scopes and meanings. Various forms of correctness may be attributed to it. The new version of the text thus makes it possible to adopt radically divergent *points of view* on the theorem, points of view that may nevertheless all appear to coincide with Elkan's own. By that very token, this de-monstration constitutes an instrument apt to help reduce the disagreements that separated the author from his challengers and to help stabilize the controversy.

The relative opacity of Elkan's formulations had consequences for a particular subset of readers (an audience of "non-specialists" in fuzzy logic, to borrow Ruspini's term) who could not fully evaluate the criticisms leveled against Elkan's de-monstration during the debates, in particular those having to do with the notion of logical equivalence. This might be explained by their relative lack of knowledge on the subject, or by the limited time they were prepared to spend reading texts by Elkan and his detractors.[22] The relatively "fuzzy" character of some of the language used in the second version of the article did not make it any easier to grasp the presuppositions and implications of a particular notion of logical equivalence. Thus certain readers were led to forge a viewpoint about Elkan's theorem on the basis of a partial decoding of the arguments put forward.

Under these conditions, references to earlier results and a confident tone were among the elements capable of winning adherence. Given the tokens of seriousness Elkan was offering, and lacking any other criteria of evaluation, a "nonspecialist" reader of his introduction could simply retain the conviction that the author's result was valid for "several" notions of logi-

---

[22] See the analyses of the exchanges on comp.ai.fuzzy, above.

cal equivalence, without judging as problematic the use of one particular notion of logical equivalence or another in fuzzy logic.[23]

Elkan also made changes at the end of the second section in his commentary on the proof itself. He partially reorganized this passage, retaining certain statements, adjusting others, and introducing new remarks.

The first innovation lies in the reference to a result by Dubois and Prade, which is presented in the November version as comparable to the one demonstrated in Elkan's own article. In the context of the electronic forum, reference to this type of result enabled critics to show that "the" theorem proposed by Elkan was not at all original. In the second version of the article, Elkan makes a completely different use of this type of reference.

The citation now constitutes a new resource to reinforce the credibility of the theorem, especially in the eyes of readers who do not have the ability or the resources (in particular, time) to assess the pertinence of such a link. In the process, the citation also complicates the task of Elkan's detractors. Indeed, it tends to constrain those who would seek to criticize his de-monstration, forcing them to position themselves with respect to the various results evoked, and thus risk losing a certain number of their own "nonspecialist" readers.

To be sure, this declaration then raises the question of the value of the proof, to the extent that the proof is no longer presented as original. A new commentary provides elements that help circumvent this difficulty. Elkan's theorem is presented as "stronger" than Dubois and Prade's result, in terms of the specificity of the hypotheses used, whereas the latter is said to be stronger than Elkan's, in terms of the general character of the hypotheses it requires:[24]

> The link between Theorem 1 and this proposition is that $\neg (A \wedge \neg B) \Leftrightarrow B \vee (\neg A \wedge \neg B)$ is a valid equivalence of Boolean algebra. Theorem 1 is stronger in that it only relies on one particular equivalence, while the proposition of Dubois and Prade is stronger in that it applies to any connectives that are truth-functional and continuous as defined in their paper.

At the price of a somewhat acrobatic formulation, Elkan's de-monstration is once again *valorized*, even as a reassuring message is delivered to certain of its detractors, two in particular. Now that this task is out of the way, the following paragraph responds once again to objections related to the introduction of the logical equivalence that was so heavily criticized on comp.ai.fuzzy:

---

[23] Let us recall that just such a critical posture had been adopted in the debates on the electronic forum.

[24] See Elkan, "Paradoxical Success," November 1993, p. 3.

The equivalence used in Theorem 1 is rather complicated, but it is plausible intuitively, and it is natural to apply in reasoning about a set of fuzzy rules, since $\neg(A \wedge \neg B)$ and $B \vee (\neg A \wedge \neg B)$ are both re-expressions of the classical implication $A \Rightarrow B$. It was chosen for this reason, but the same result can also be proved using many other ostensibly reasonable logical equivalences.[25]

Let us recall that, for researchers such as Ruspini, the use of the logical equivalence in question amounted to calling "implicitly" on the principle of the excluded middle. As a result, certain protagonists in the debates on the electronic forum had represented Elkan's result as trivial.

Without referring to this criticism, Elkan justifies his recourse to this particular logical equivalence by introducing a new de-monstrative register. New evaluation criteria are applied to the equivalence, which is characterized as "rather complicated," but also as "natural" and "intuitive."

In this passage, Elkan is using a reference to the realm of intuition to justify a formal choice. His approach can be compared to the one we have already examined in Husserl. It brings to light once more the alternative between approaches that we have characterized respectively as logicist and formalist: the self-evidence of a logical equivalence may be invoked without appealing to a notion of formal coherence (by attributing to the equivalence an "intuitive" or "natural" character, for example), or by appealing to this notion alone.

The use of the equivalence that triggered so many objections is thus legitimized without the adoption of an overly defensive stance. Moreover, the allusion to the possibility of achieving the same outcome with "several other" "apparently reasonable" notions of logical equivalence constitutes an element apt to make the theorem appear very general in the eyes of readers who are more or less incapable of grasping the technical aspects of the argument.

After introducing these new de-monstrative registers, Elkan then takes up certain comments from the first version of his article and modifies them to varying degrees. To begin with, at the price of some minor corrections, he insists once again on the following points: the validity of his logical de-monstration does not depend on the definition of logical implication retained; his result extends to any more general formal system that includes the four postulates mentioned at the beginning of his de-monstration;[26] his theorem also applies to fuzzy set theory as well as to other systems of fuzzy logic developed by other authors that use other logical operators for negation, disjunction, and conjunction.

---

[25] Ibid.

[26] Elkan is referring here to the logical systems that mobilize not only the four postulates in question but also other axioms that are formally compatible with the preceding ones.

The theorem's "broad generality" is displayed yet again, at least in the eyes of readers apt to be receptive to this interpretation. However, Elkan's comments may also respond to a series of specific objections that the author had to face even before the publication of his article in AAAI'93. Let us recall that Elkan had invited some of his colleagues to look at his text when it was only in draft form. Moreover, we find a trace of such a step in the second paragraph of both versions of the article, which is devoted to the author's acknowledgments. Such a hypothesis allows us in any event to account for the evolution of the theses formulated in this passage in relation to those Elkan defended during my first interview with him. In drafting his own critiques, the author did not in fact initially intend to take into account a plurality of axiomatics associated with fuzzy logic.

In the last paragraph of the second version of his text, Elkan betrays his awareness of cases for which his theorem "does not hold," even as he makes a point of their limited importance:

In order to preserve a continuum of degrees of truth, one naturally wants to restrict the notion of logical equivalence. In intuitive descriptions, fuzzy logic is often characterized as arising from the rejection of the law of excluded middle: the assertion $A \vee \neg A$. Unfortunately, rejecting this law is not sufficient to avoid collapse to just two truth values. Intuitionistic logic rejects the law of excluded middle, but the formal system of Definition 1 still collapses when logical equivalence means intuitionistic equivalence.[27] Of course, collapse to two truth values is avoided when one only admits the equivalences generated by the operators minimum, maximum, and complement to one. However, these equivalences are just the axioms of de Morgan, which only allow restricted reasoning about collections of fuzzy assertions.[28]

In the first sentence of this paragraph, Elkan shows that he is not unaware of the constraints imposed on a formal system by the choice of a particular notion of logical equivalence. One might suppose that this assertion follows directly from the criticisms expressed on the subject on the electronic

[27] At this point, Elkan introduces the following note: "The Gödel translations [Dirk van Dalen, *Logic and Structure* (New York: Springer-Verlag, 1983), p. 172] of classically equivalent sentences are intuitionistically equivalent. For any sentence, the first three postulates of Definition 1 make its degree of truth and the degrees of truth of its Gödel translation equal. Thus the proof in the appendix can be carried over directly. Dubois and Prade ["An Introduction to Possibilistic and Fuzzy Logics," in *Non-standard Logics for Automated Reasoning*, ed. Philippe Smets et al. (New York: Academic Press, 1988)] note that if all the properties of a Boolean algebra are preserved except for the law of excluded middle, their proposition no longer holds. This observation is compatible with a collapse assuming only the equivalences of intuitionistic logic, because although intuitionistic logic rejects the law of excluded middle, it admits a doubly negated version of the law, namely $\neg \neg (\neg \neg A \vee A \neg A)$."

[28] Elkan, "Paradoxical Success," November 1993, p. 4.

forum. However, it is found in a very similar form in the first version of the text. It is thus possible that Elkan had had to confront analogous criticisms (even though these latter were not as well supported) before the first publication of his article.

In contrast, Elkan addresses the principle of the excluded middle and its rejection in fuzzy logic in new terms. In his first version, Elkan made the following assertion: "What all formal fuzzy logics have in common is that they reject at least one classical tautology, namely the law of excluded middle (the assertion $\neg A \vee A$)." In the second version, he offers a response to the criticism formulated in particular by Ruspini on comp.ai .fuzzy. Elkan no longer asserts that all systems of fuzzy logic reject the principle of the excluded middle. To this end, the realm of "intuition" is once again brought into play. Elkan associates the rejection of this principle in fuzzy logic with "intuitive descriptions" that are "frequently" made by *third parties*. Moreover, he does not mention the thesis according to which the equivalence that has so often been criticized *amounts* to introducing the principle of the excluded middle.

This formulation could lead a certain number of readers not to see a limitation on the value of the theorem here. However, it also could lead readers who had followed the debates to think that a "gross error" had not been made, the one that would have consisted in unwittingly introducing the principle of the excluded middle.

On the basis of the foregoing excerpt from the first version of the article, one may wonder how Elkan could have realized, before his text was delivered in July 1993, that the equivalence so often criticized on the electronic forum could lead to the introduction of the principle of the excluded middle. Thus we understand better, a posteriori, Ruspini's remark on this subject on comp.ai.fuzzy.[29]

However, what follows in the paragraph cited above tends to present Elkan's theorem as not without value, even though it might rely on the "implicit" introduction of the principle of the excluded middle. Indeed, Elkan claims that his result remains correct when the notion of equivalence used in intuitionist logic is adopted, whereas that logical system rejects the principle of the excluded middle. For readers unfamiliar with these notions, Elkan's theorem could thus appear as a wholly general result.

This thesis was already present in a slightly different form in the first version of Elkan's article. But the passage in question seems to have

---

[29] See again Enrique Ruspini, message 816: "One can only wonder to what kind of analysis, if any, this claim was subjected by the NCAI [National Conference on Artificial Intelligence] referees, when the author claims, in the discussion following his theorem, that 'what all formal fuzzy logics have in common is that they reject at least one classical tautology, namely, the law of the excluded middle,' while basing his main argument on an equivalent assertion."

been limited to exhibiting a form of generality of the result. In the second version, it may well have had an additional function, as a response to the interventions that had pointed out and criticized the introduction of the principle of the excluded middle in Elkan's de-monstration of the theorem.

The author's corrections to the paragraph that followed include other responses to some of the criticisms expressed on the electronic forum. One objection had to do with the validity of the result in intuitionist logic.

This time, Elkan specifies in a note—a new element—that intuitionist logic admits a "version" of the principle of the excluded middle (in the form of a double negation of this principle). This statement gives readers familiar with these questions a picture of an author who is aware of the validity in intuitionist logic of a statement making it possible to demonstrate the same result rapidly (by relying in particular on the third of the four postulates evoked above). At the same time, placing this "remark" at the end of a note without further commentary helps to avoid devaluing the de-monstration in the eyes of readers who have not followed the debates.

Elkan presumably attached a certain importance to the criticisms that had been addressed to him on this issue. This hypothesis would explain why he proceeds to thank unnamed persons in the second version of his text, as if he did not want to "embarrass" any colleague to whom he had submitted his drafts: "The author is grateful to many colleagues for useful comments on earlier versions of this paper."[30] In the first version, he thanked a certain John Lamping, who had asked him whether his theorem still "held up"[31] when the notion of equivalence was understood intuitionistically: "The author is grateful to several colleagues for useful comments on earlier versions of this paper, and to John Lamping for asking if theorem 1 holds when equivalence is understood intuitionistically."[32]

The last part of the paragraph under study reveals an author aware of the fact that the theorem would "of course" no longer be correct if one assumed only the first three axioms used in his de-monstration. This point was not made in the first version of the article. Elkan is thus responding to the criticisms expressed on this point on the electronic forum. However, he notes the reduced importance of this case by declaring that logical

[30] Elkan, "Paradoxical Success," November 1993, p. 12.

[31] The protagonists' declarations about what makes a theorem "hold up" are particularly useful elements to be analyzed in order to grasp the establishment of priorities in the exercise of logic, and to see what makes the proofs' steps appear necessary for the actors.

[32] Elkan, "Paradoxical Success," August 1993, p. 702. See J. O. Lamping, "A Unified System of Parametrization for Programming Language," Ph.D. diss., Stanford University, 1988.

equivalences that depend on the first three axioms alone would allow only "limited" reasoning about fuzzy statements.

This remark is formulated differently from the corresponding remark in the first version; it clearly does not play the same role. In the first version, Elkan declared that he doubted the existence of a notion of logical equivalence that would have both a "practical" value and a "philosophical" ground, and that would not make it possible to obtain a result "analogous" to his own: "It is an open question how to choose a notion of logical equivalence that simultaneously (i) remains philosophically justifiable, (ii) allows useful inferences in practice, and (iii) removes the opportunity to prove results similar to theorem 1."[33]

In this first version, the cases for which the theorem was found wanting were thus presented a priori as marginal. A larger number of registers were used on this occasion, corresponding to the diversity of commitments and skills that characterized the author's trajectory and his practice of research. But the adoption of a utilitarist and philosophist position (if these terms can be used to characterize the idea that philosophy "in general" or practice "in general" would "impose" some particular notion of logical equivalence) presumably gave rise to criticisms that led Elkan to abandon such an assertion.

In sum, we see how the restructuring of the comments accompanying Elkan's proof and the production of new remarks in the second version of the article offer answers to the objections expressed on the electronic forum. They put on display an author who was not a victim of "gross errors" and a theorem of entirely general scope in the eyes of "nonspecialists." For these latter, the citations of works by detractors of Elkan's article tend to reinforce the credibility of the result, while they also limit the maneuvering room available to Elkan's critics in the continuation of the debates.

Certain formulations, by virtue of their allusive and polysemic character, seem capable of delivering different messages to different sets of readers. These formulations, aimed at reassuring the most critical researchers in fuzzy logic, are also designed to assuage the concerns of the author's previous supporters. Displaying formulations that may appear more "modest" and more "accurate" to the first group, the new text is simultaneously positioned to display a "strong" and "general" result to the second group.

The vagueness of certain expressions and the reader's need to have highly specialized skills to grasp the exact scope and also the limits of certain statements are not the only factors at work here. Other devices convey specific arguments to various types of readers (depending on

---

[33] Elkan, "Paradoxical Success," August 1993, p. 699.

whether or not they are in a hurry, how much they care about the debates, what skills they have, and so on): the diversity of the formulations retained from one passage to the other (which in certain cases may appear contradictory to one group of readers), and also for example the variable degrees of visibility of the paragraphs in which they are included.

During an interview with me in early 1994, John Carpenter declared, not without irony, that he found it "astonishing" from a sociological standpoint that Elkan had become aware of the "error" that he had made (by introducing the equivalent of the principle of the excluded middle in his proof) even as he "persisted in the error" in his public interventions. The foregoing descriptions show that such a dynamics does indeed deserve to be subjected to sociological analysis. The analysis we have just undertaken makes it possible to attribute multiple meanings to Elkan's text instead of judging it "meaningless." Owing to this approach, we have not had to seek in—or attribute to—Elkan's interventions any unity of thesis or voice (however surprising this may seem insofar as the deployment of logical results is concerned).

This approach turns out to have another fundamental virtue. It has allowed us to understand how Elkan's new and improved textual device could have had at least two effects: it limited the conflict that opposed him to his challengers, and it rallied to his theses readers who ultimately adopted hostile views toward "his" theorem, in particular attributing to it various forms of correctness.

It is nevertheless important to stress the quite specific form of rallying in question here. I am not asserting that Elkan managed to construct *a* "stronger" position than the others, sublimating all the other positions while integrating all the participants' objections. I am not claiming that the author succeeded in winning the effective and unconditional adherence of all these readers to the new text he produced. I am describing quite a different situation. The new version of the text could limit the criticisms produced by Elkan's adversaries and convince some of his readers of the "correctness" of his theorem: such readers could agree on this point even while adopting antagonistic viewpoints on the nature of the theorem thus de-monstrated and on the meaning to be attributed to the term "correctness." These two outcomes could lead to a stabilization of the controversy.

Moreover, I do not claim that any one reader perceived the multiplicity of the meanings that could be assigned to Elkan's undertaking. The analysis of Carpenter's last interview illustrates this. However, Carpenter's representation must be understood in the light of his particular investment in the debates and the entire set of competences available to him. Such a configuration was highly exceptional.

It is also important to emphasize that, for Elkan, the polysemic character of the reformulations represented an important tool for continuing the debates. With the support of his new text, the author had maneuvering room to propose personalized and variable presentations of his theses to very different interlocutors, without much risk that he would be accused of inconsistency. Presumably, without these margins for maneuver, the tailor-made responses that Elkan developed starting in November 1993 would be hard to imagine.

The analysis of the two versions of the article allows us to identify possible links between the alterations made in the text and the evolution (or even the stabilization) of the controversy. In addition, it brings to light resources that are clearly essential to the debates, and shows how the simple rewriting of a text can constitute an incomparable tool for answering a multitude of criticisms.

This study extends Warwick's analyses of the conditions for the emergence of the theory of relativity.[34] Warwick showed how the specific graphic methods used played a role in the rise and stabilization of that theory. For the sociological analysis of the stabilization of a scientific statement, we can now see that it may be useful not to limit oneself a priori to the description of *graphic* processes, but to develop the tools necessary for a detailed study of the *symbolic* manipulations in play.

This does not mean that one has to adopt a semiotic approach that attributes univocal meanings to the signs used by the actors, even if these signs are symbolic in nature. We would never have noticed the process under study if we had been content to try to pin down the construction of a single reader in his or her relation to symbols. We would simply not have seen how several readings of the same signs were possible, given the variable resources and skills at the protagonists' disposal and the dynamics in which their confrontations with the texts were inscribed.

For us to grasp such a process, we ultimately have to see Elkan's readers less as ideal readers of certain works in the semiotics of mathematics than as figures in Proustian descriptions. Proust's fictional characters are often confronted with polysemic formulations that can convey distinctly different messages to different actors. And these actors indeed have very different ways of apprehending signs that might seem most likely to support a single interpretation and a programmed effect.

A posteriori, we note the extent to which I would have been ill advised to limit my investigations to Elkan's texts alone. And yet that seemed to be the obvious path. Let us recall that Livingston focused his study on Gödel's incompleteness proof.[35] Pickering and Stephanides, for their part,

---

[34] Warwick, "Cambridge Mathematics and Cavendish Physics," part 1.
[35] Livingston, *Ethnomethodological Foundations of Mathematics.*

sought to account for a research activity that was, a priori, comparable to Elkan's, and they examined very few documents.[36] Of course, the goal of these sociologists lay elsewhere, and their approach was fruitful. But my point, too, lies elsewhere: if I had adopted such an approach, I would not have been able to discover how the smallest details of the successive versions of Elkan's de-monstration were linked to the author's prior inter-actions and to the anticipation of future relations with third parties, and how they at once helped make future exchanges possible and partially constrained the content of such exchanges.

At this stage, we can appreciate the nature of the task that would still have to be accomplished if we were to analyze the corrections Elkan made to the rest of his article. However, it is not necessary to go further along this path, for the study of the first two sections has brought to light the procedures that characterize the transformations made to the first version of the text. In contrast, it is now important to show that in order for a proof like Elkan's to be the object of a revised formulation, it did not have to give rise to a controversy first. The author had already made corrections to his de-monstration even before it was published. Such ad-justments stemmed more generally from the submission of the text of the proof to critical readings by the author himself and by third parties.

To make such phenomena clear, we can compare the content of the article published in the proceedings of AAAI'93 with that of a draft from early 1992. Let us recall that, at that time, Elkan had given me copies of several of his texts, some not yet published, in order to make my study possible. The draft in question was part of this set.

The Successive Versions of a Proof: Records of Negotiations

When we compare the draft of the paper Elkan gave me in San Diego in 1992 with the text published in AAAI'93, we observe a very considerable evolution. The title of the early draft, "Probability vs. Fuzziness," already announced a very different project. In this preliminary version, Elkan sought to refute certain theses formulated by Kosko in an article titled "Fuzziness vs. Probability."[37] Let us recall that, by Elkan's own account, Kosko's was one of the first papers he had read on fuzzy logic.

In the introduction to the early draft, Elkan presented his main objective: to analyze "the presuppositions of fuzzy logic" in the face of Kosko's criti-cisms of probability theory. The article had several sections, beginning with

[36] Pickering and Stephanides, "Constructing Quaternions."
[37] Kosko, "Fuzziness vs. Probability."

a general reflection on what made a "system of laws of thought" valuable. In this way Elkan introduced a discussion of the respective properties of probability theory and fuzzy logic in relation to various forms of uncertainty. After comparing the way fuzzy logic and probability theory dealt with a "small problem," that of determining the color of a collection of watermelons (were they green inside, or red?), the author concluded that, unlike probability theory, fuzzy logic was unable to take uncertainty into account.

The introduction announced that the following section would focus on a particular point: the question of whether fuzzy logic actually made it possible to resolve Russell's paradox (as Kosko claimed in his article). Without making an explicit link with this problematics, Elkan then developed a version of the proof we have examined here. The de-monstration ended up with a different result, since it led to the assertion that fuzzy logic was inconsistent.

The author then proposed to show that, contrary to Kosko's claim, fuzzy logic did not constitute a generalization of probability theory. This version of the article did not include other developments, or even a conclusion, although in his introduction Elkan announced that he was going to devote a final section to other issues that had not been examined in the previous sections.

A glance at the overall structure of this draft shows that it actually has little in common with the text published in the AAAI'93 proceedings. To be sure, we find a great variety of de-monstrative registers in both cases. But only the symbolic de-monstration and the evocation of the "small problem" of assessing the color of a collection of watermelons are taken up again in the August 1993 version. The other developments, particularly the considerations stemming from general philosophy and most of the theses on probability theory, have been abandoned. Bart Kosko, whose article was at the heart of Elkan's earlier reflection, no longer constitutes either a central interlocutor or an explicit spokesperson for fuzzy logic in the August 1993 version.

The main arguments of the presentation apparently did not stand up to the critical readings mentioned above. Elkan's revision of his earlier theses seems to correspond to a desire to reply to a growing number of objections and self-criticisms, a desire to consolidate his de-monstration even if this meant modifying his object (for example by addressing the question of the properties of fuzzy logic through an analysis of the way its applications worked). This evolution is highlighted by the extension of the bibliography (from six to thirty-three references) and the transformations made in the symbolic de-monstration.

The first version of the proof, as we have seen, ends up with a very different result: Elkan de-monstrated that fuzzy logic had *only one* truth value. Thus he asserted that it was inconsistent. Starting from the statement of the four axioms that would remain associated with fuzzy logic in the later versions of his article, the author introduced his de-monstration with commentary in three phases. He specified that the expression "logically equivalent," used for the definition of the fourth axiom, signified "equivalent" according to the rules of standard propositional calculus with two truth values. Raising the question whether the operations of fuzzy logic succeeded in grasping the combinatorial behavior of truth values, he added that he would consider that a logical system to be inconsistent by definition if it assigned the same truth value to all statements.

At this stage, Elkan took fewer precautions, and he did not announce, as he would in the August 1993 version, that he was going to comment on his choice of the notion of logical equivalence at the conclusion of his presentation of his de-monstration. It seems that he had not yet had to confront objections on this point. This led him both to affirm that fuzzy logic was inconsistent, and to posit this statement as a theorem that he was going to demonstrate.

The goal of Elkan's proof was not, at this point, to show that fuzzy logic *in fact* constituted a binary logical system. Even less was the author seeking to de-monstrate that the introduction of an "apparently reasonable" logical equivalence led to inconsistencies in fuzzy logic.[38] What Elkan was attempting to prove at this stage is that fuzzy logic *was* inconsistent. Now how could the author have de-monstrated in his draft that fuzzy logic possessed only one truth value, whereas several months later he showed that it had two?

Although this first form of the statement of the theorem was very different from the one included in the August 1993 text, the two versions of the symbolic proof were very similar. The author started with the same logical equivalence, used the same calculus, and obtained the same list of seven double numerical inequalities. The only correction Elkan made to the text of his proof came in the very last paragraph. In the August 1993 article, Elkan concluded his proof as follows:

Now let $x = \min\{t(A), 1 - t(A)\}$ and let $y = \min\{t(B), 1 - t(B)\}$. Clearly $x \leq 0.5$ and $y \leq 0.5$ so if not $x = y$, then one of the eight inequalities derived must be satisfied. Thus $t(B) = t(A)$ or $t(B) = 1 - t(A)$.[39]

In his draft, the author ended his de-monstration by affirming simply that "only if $t(A) = t(B) = 0.5$ is it possible for none of the eight inequalities derived to be satisfied." When we compare these two versions, it appears that the critical readings (including self-critical ones) to which the article was submitted before publication defeated that last step in the de-mon-

[38] Let us recall that this is one of the possible readings of the November 1993 version of the article.
[39] Elkan, "Paradoxical Success," August 1993, p. 699.

stration. Let us recall that in the terminology adopted here, a *step* constitutes a *monstrative* moment, a moment in which the author was content to *show*, (re-)defining by that very token a borderline between what could and could not be shown as such.

Something that represented a correct step in the demonstration for the author, something that did not call for further justification and that could be simply shown as such, became the object of unexpected viewpoints. Other justifications then *appeared* necessary. The *questioning* of the last de-monstrative step led to a new *shaping*, including a set of supplementary justifications and other de-monstrative *steps*, and this led in turn to a shift in the result. The need to de-monstrate what was initially presented as self-evident led to a negotiation and a transformation of the object of the de-monstration.

We see here how Elkan's proof could be consolidated at the price of new manipulations, the affirmation of the need for a new series of demonstrative steps, and, correlatively, a transformation of the theorem. While the two versions of the statement of the theorem appear profoundly different by virtue of the simple modification of the last phase of the demonstration, the same holds true for the considerations that follow the presentation of the proof. Here again, the author took pains to transform the brief commentary included in the draft into a set of carefully considered remarks. He devoted four long paragraphs to commentary on the proof in the August 1993 text, whereas he was content with four sentences in the 1992 draft. In this earlier version, he asserted that fuzzy logic was not a fully specified formal system and that, in particular, no rules concerning any implication connective were stated. He added that the theorem did not depend on the definition of implication as $\neg (A \wedge \neg B)$, and that any formal system that included the four axioms stated was inconsistent.

The critique of fuzzy logic was much more radical here than in the later version. But it was also much less well defended. Both more direct and less nuanced, it did not rely on any references. Elkan had apparently not yet integrated the objections that could shift the object of his de-monstration. In particular, he did not attempt to show, as he would later on, that his theorem remained correct for other notions of logical equivalence.

The successive versions of Elkan's article thus seem to retrain traces of the influence of the critical readings to which the author referred in his article or in his interviews with me. Elkan's de-monstrative production seems to be inscribed in cycles of proofs and refutations comparable to those described by Lakatos in the case of Euler's theorem on polyhedra.[40] The different versions of Elkan's text form a record of negotiations. As

---

[40] See again Lakatos, *Proofs and Refutations*, and Bloor's analysis of the controversy over Euler's theorem in "Polyhedra" (1982).

objections arose (and some of these may have come from the author himself), the proof problematized more and more de-monstrative steps and logical notions. It became more resistant because less direct, more nuanced and better accompanied. The theorem, the proof, and the logical objects incorporated (like the notion of logical equivalence) were transformed *simultaneously*. From one version to the next, certain elements of the proof were abandoned, others were brought to light and exploited in new de-monstrative sequences. By following the steps in the de-monstrative production, we observe that the work of proving put into play *skillful bricolages*.

This allows a better understanding of the thesis advanced by De Millo, Lipton, and Perlis, who stressed the crucial role played by interactions in the process of rewriting proofs. According to these researchers, such dynamics reinforce the power of de-monstrations to attract attention and to "convince."

While each version of Elkan's de-monstration may appear definitive, this does not mean that the text is a fixed entity. Integrating criticisms as they arose, the successive shapings of the article aimed to produce a more and more resistant result.

Moreover, the launching of a controversy was not a prerequisite for correcting the de-monstration. As we have seen, Elkan made important modifications to the draft of his article before its publication in the AAAI'93 proceedings—that is, before the debates began.

We are now ready to determine whether the conclusions of these analyses are confirmed by the study of texts produced starting in 1994, when the controversy unfolded in several journals. In particular, while in this chapter we have focused exclusively thus far on Elkan's own production, we also need to examine the responses of researchers in fuzzy logic. Will the analysis of their texts allow us to distinguish other aspects of the work of proof? What de-monstrative strategies do these texts put to work in confronting Elkan's various articles and papers, and what are the consequences? Finally, we need to look at the outcome of this process. Does the controversy end up being stabilized, and does Elkan's theorem end up achieving the status of certified knowledge (which still has to be defined)?

DE-MONSTRATIONS SERVING TO STABILIZE A CONTROVERSY

I have sought to identify all the publications devoted to the discussions of Elkan's theses systematically, from 1994 on. My investigations have shown that these latter were being cited and to some extent disputed[41]

---

[41] See in particular Fred A. Watkins, "False Controversy: Fuzzy and Non-Fuzzy Faux Pas," *IEEE Expert*, April 1995, pp. 4–5; Xiang Zhao Shu and Ping Lu Jian, "Measure-

right up to the time the present book went to press.[42] Still, the controversy seems to have nearly petered out in 1994 after the publication of a series of articles that deserve mention: two in *AI Magazine*,[43] another coauthored by Dubois and Prade in the proceedings of the AAAI'94 conference,[44] and finally a special issue of *IEEE Expert* entirely devoted to continuing the debates.[45]

The appearance of the special issue of *IEEE Expert* in August 1994 (nearly a year after the first publication of Elkan's article in the AAAI'93 proceedings) seems to have represented a high point in the debates. This observation does not mean of course that a reopening of the controversy is excluded, even after the appearance of the present book.

Among the articles cited above, those published in 1994 in *AI Magazine*, along with one of the sections in comp.ai.fuzzy's FAQ from May 1994, warrant particular attention; the latter in fact proposes to recapitulate the arguments exchanged during the debates.

Federating and Stabilizing Positions and Thereby Helping to Marginalize the Adversary

The section reserved for correspondence from readers ("Members' Forum") in the spring 1994 issue of *AI Magazine* included two texts de-

---

Based Fuzzy Operators Explain Elkan's Theorem," *IEEE Expert*, January 1996, pp. 84–85; Juiyao Pan, Guilherme N. DeSouza, and Avinash C. Kak, "FuzzyShell: A Large-Scale Expert System Shell Using Fuzzy Logic for Uncertainty Reasoning," *IEEE Transactions on Fuzzy Systems* 6, no. 4 (Nov. 1998): 563–81; Claudio Sossai, Gaetano Chemello, and Alessandro Saffiotti, "A Note on the Role of Logic in Fuzzy Logic Controllers," *Proceedings of the Eighth International Conference on Information Processing and Management of Uncertainty in Knowledge-Based Systems* (Madrid, 2000), pp. 717–23; Enric Trillas and Claudi Alsina, "Elkan's Theoretical Argument, Reconsidered" *International Journal of Approximate Reasoning* 26, no. 2 (2001): 145–52; Charles Elkan, "Paradoxes of Fuzzy Logic, Revisited: Discussion," *International Journal of Approximate Reasoning* 26, no. 2 (2001): 153–55; Enric Trillas and Claudi Alsina, "Comments to 'Paradoxes of Fuzzy Logic, Revisited: Discussion,' " *International Journal of Approximate Reasoning* 26, no. 2 (2001): 157–59; Guo Jun Wang and Hao Wang, "Non-fuzzy Versions of Fuzzy Reasoning in Classical Logics," *Information Sciences* 138 (Oct. 2001): 211–36; Carl W. Entemann, "Fuzzy Logic: Misconceptions and Clarifications," *Artificial Intelligence Review* 17, no. 1 (2002): 65–84; Paul Snow, "What Concrete Things Does Fuzzy Propositional Logic Describe?" Eighth Pacific Rim International Conference on Artificial Intelligence, Auckland, New Zealand, August 9–13, 2004, *Proceedings Lecture Notes in Artificial Intelligence*, vol. 3157 (Berlin: Springer, 2004), pp. 445–53; Radim Belohlavek and George J. Klir, "On Elkan's Theorems: Clarifying Their Meaning Via Simple Proofs," *International Journal of Intelligent Systems* 22, no. 2 (2007): 203–7.

[42] A few messages prolonging the exchanges also appeared on comp.ai.fuzzy.

[43] *AI Magazine*, Spring 1994, pp. 6–8.

[44] Dubois and Prade, "Can We Enforce Full Compositionality?"

[45] *IEEE Expert*, August 1994.

voted to the continuation of the debates.[46] The first is presented as a critique of the article Elkan published in the AAAI'93 proceedings. The second, by Elkan himself, offered a new response to the objections addressed to his theses.

The first document, "A Reply to the Paradoxical Success of Fuzzy Logic," was signed by eight people, prominent figures in fuzzy logic or mathematicians working on various theories of uncertainty. Elkan's article was thus subjected to the test of numbers. The reader became a witness to the consensus that had arisen around the denunciation of Elkan's theses, at least so far as several leading figures in fuzzy logic were concerned. One of the questions "implicitly" formulated was the following: if these eight researchers all agreed to denounce the same errors, how could Elkan not be wrong?

The cosigning in itself thus had a properly de-monstrative function, even if it was perhaps more incidental than intentional. It could have been the result of editorial constraints, a limitation on the volume of interventions imposed by the journal.[47] One may even speculate that the authors *had* to come to agreement on a common text in order to be published.

However, one may also speculate that, for the most part, only certain of the signers drafted the text; the others may have been content to offer comments or simply to co-sign it in order to give it additional legitimacy. In fact, the style of the article is perceptibly different from that of the message coauthored by Dubois and Prade for the electronic forum.[48]

However this may be, we see here how the transposition of the debates into a new forum contributed to a stabilization of the controversy. Whereas a multitude of viewpoints had been expressed on comp.ai.fuzzy, editorial constraints turned the debates into a head-to-head confrontation between Elkan and a group serving as spokesperson for fuzzy logic.

This reduction in the number of viewpoints was equaled only by the stabilization of "the" response by researchers in fuzzy logic. After a short introduction, the authors focused on bringing to light three errors in Elkan's text. The first, according to them, involved the symbolic proof; the second occurred in the treatment of the small problem of estimating the color of a collection of melons; the third arose in the analysis of the way fuzzy controllers work. In their conclusion as well as in their introduction, the authors justified their efforts by spelling out the risks that Elkan's theses entailed for fuzzy logic:

[46] *AI Magazine*, Spring 1994, pp. 6–8.
[47] Let us recall that the response was published in the section devoted to letters from readers.
[48] Of course, the hypothesis that ideas were shared during a meeting prior to the drafting of the article or that some other form of concertation led to the attribution of mandates to the author (or authors) of the text cannot be ruled out either.

A reply to the author's comments seems to be necessary in order to stop poten-tially deleterious effects of such a paper and to avoid further propagation of such fallacies in the scientific community. . . . The problem with such papers is that casual readers will use them to criticize and maybe stop future fundings for research on the topic, just as it was the case with the famous British report that killed artificial intelligence for a decade in Great Britain.[49]

The authors of this article were not the first to present Elkan's text as a threat to fuzzy logic and its continued funding. Ruspini in particular had evoked such risks in his message on the electronic forum. However, the passage cited here developed the point by recalling an event capable of making an impression on researchers in "classical" artificial intelligence: a drastic budget cut for this field of research in the United Kingdom, repre-sented as a "consequence" of a devastating report.[50]

This article drew on another de-monstrative register used by Ruspini, the denunciation of "lack of care" on the part of the AAAI'93 reviewers:

The fact that such a paper was accepted by the Scientific Committee of AAAI is really not in favor of the reviewer's work! Their selection deserves greater care. . . . Acceptance and publication of negative papers that claim to "prove" that a theory is wrong should be made with more care.[51]

In fact, the entire set of objections developed by the eight authors did not go very far beyond those formulated by Ruspini in his message on comp.ai.fuzzy, as an examination of the criticisms addressed to Elkan's symbolic de-monstration will show. A summary of their argument follows.

The result achieved by Elkan, given the axiomatics adopted, has been known for a long time. To demonstrate this, it suffices to carry out a simple calculus, starting with the statement of the principle of the excluded middle. But fuzzy logic cannot be defined by Elkan's axio-matics. The classical notion of logical equivalence does not hold in fuzzy logic. In particular, the logical equivalence between the propositions "$(\neg A \wedge \neg B)$" and "$(\neg A \wedge \neg B) \vee B$," introduced by Elkan, is not applicable to fuzzy propositions. Consequently, it is Elkan's proof that "collapses," and not fuzzy logic. Elkan's reasoning does not hold in intuitionist logic, either, to the extent that the author presupposes that negation is involutive.[52]

---

[49] Berenji et al., "Reply to Paradoxical Success," pp. 6, 8.

[50] During my study, I observed that this episode was indeed present in the minds (or rather in the discourse) of a number of researchers in artificial intelligence in the United States.

[51] Berenji et al., "Reply to Paradoxical Success," pp. 6, 8.

[52] To presuppose that "negation is involutive" is to make the following hypothesis: the double negation of any proposition "P" is "equal" to that same proposition "P." In other words, "not (not P)" is "equal" to "P."

These criticisms constitute a faithful restatement of the objections Ruspini formulated on comp.ai.fuzzy. While the electronic forum presented an explosion of divergent opinions about Elkan's de-monstration, we now see the criticism of this proof in the process of being stabilized from two different standpoints. On the one hand, a single judgment rather than a multitude was expressed. On the other hand, the content of the criticism was established in fixed form through the use of de-monstrative registers already brought to bear by Ruspini.

The response developed by Ruspini thus offered researchers in fuzzy logic a support structure for the development of a de-monstrative response based on the repetition of the same arguments in diverse forums. Such an undertaking is situated at the opposite pole from the one that Elkan had adopted up to that point. Elkan had devoted time and energy to renewing and adjusting his formulations, which tended to become more and more polysemic. In comparison, the approach of the prominent voices in fuzzy logic appeared less "costly." It tended to supply the image of a "community" united around a coherent answer, unlike the contradictory responses that appeared on the electronic forum.

This consistency was manifested just as much in the refutations of Elkan's other theses, including those bearing upon the operation of fuzzy controllers, as it was in the various challenges to his formal result. Thus to Elkan's assertion that the applications of fuzzy logic developed up to that point were not very sophisticated *because* they used a small number of rules, the eight authors formulated a response analogous to Ruspini's. They stressed the specific *advantage* of the applications of fuzzy logic in being able to function with a relatively small number of rules. They went further than Ruspini in developing the argument, however. In particular, the authors denounced Elkan's lack of understanding of the differences between the "approach" of the inventors of classical expert systems and that of the originators of fuzzy controllers.

Similarly, the collective article took up Ruspini's criticism of Elkan's proposed treatment of the watermelon problem and developed it further—an important step, for the watermelon question had not been addressed to any significant degree on the electronic forum. The study of Elkan's arguments allowed the authors to assert once again that Elkan had not "understood" fuzzy logic. Berenji and his colleagues showed that Elkan had not managed to pin down the methods offered for dealing with uncertainty by probability theory and fuzzy logic respectively.

The response mode adopted by the eight authors thus consisted above all in formulating critiques analogous to Ruspini's. It tended to *marginalize* Elkan's viewpoint. Elkan's theses were submitted in effect to the test of numbers, to a stable and univocal critique, as well as to various narrative devices. Beyond those already mentioned, let us also note the recourse to *markers* such as those present in a sentence of the conclusion (the use of bold type by the authors tends to marginalize Elkan's position even further):

> Unfortunately, the author falls in a classical trap. He considers a theory, presents **his** interpretation, shows the inadequacy of the conclusions derived from **his** interpretation, and concludes with the inadequacy of the theory. He forgets that the inadequacy might reside in **his** interpretation, as was the case in the present analysis of fuzzy logic.[53]

A look at Elkan's response to these criticisms reveals, first of all, that his text was much shorter than that of his detractors: it consisted in just three paragraphs. (It is likely that the length of the response was limited, here again, by editorial constraints.) The title supplied was nothing but an excerpt from a sentence in the body of the text: "The theorem in my paper is correct . . ." This title stood in opposition to the one that had been attributed in the same fashion to the text of the eight authors: "The author fails to acknowledge that by definition, classical logical equivalence does not apply to fuzzy assertion." This sentence was printed in large type right in the middle of the article.

This graphic device helped create a particular image of the debate while also helping to stabilize it. The debate seemed to stem from a simple head-to-head confrontation, centered on the question of the "correctness" of Elkan's proof: the author maintained that his theorem was correct, whereas the researchers in fuzzy logic asserted that it was erroneous. The importance of the symbolic proof in Elkan's general critique of fuzzy logic was emphasized and reinforced.

However, Elkan's response in *AI Magazine* was hardly in keeping with such an image. Two of the three paragraphs were devoted to other points. The author asserted that a deceptive picture of the "real" content of his critique had been presented. Evoking a straightforward hostility to his theses, he invited the readers to look at his AAAI'93 paper and at the new version about to come out in *IEEE Expert*:

> I regret the hostile tone of the previous letter, which gives a misleading picture of my paper. I urge interested readers to draw their own conclusions after reading the original paper. . . . For further clarification and discussion, see a forthcoming issue of *IEEE Expert*, where the final version of my paper will be published along with responses from a number of leading researchers, including several of the signatories of the previous letter.[54]

---

[53] Berenji et al., "Reply to Paradoxical Success," p. 8.

[54] Charles Elkan, "The Theorem in My Paper Is Correct . . . ," *AI Magazine*, Spring 1994, p. 8.

An examination of this passage highlights the usefulness for Elkan, during the first half of 1994, of the version of his article that was downloadable as of November 1993 and intended for publication in *IEEE Expert*. This text served as a device on which the author could subsequently rely during his various interactions and interventions, especially when he lacked the time or space to develop long commentaries. The foregoing passage makes clear both the author's confidence in the resistance of his new text (let us recall among other things that certain polysemic formulations were apt to "satisfy" a wide range of readers) and the role played by that text in the debates from early 1994 on.

Elkan was thus in a position to do without new responses to a mass audience, and to avoid new adjustments. The result was a stabilization of Elkan's "public" positions (however polysemic they might be), as compared to those that had been accessible on the electronic forum since November 1993. An analysis of some of the commentaries Elkan published in *AI Magazine* about his own proof makes this evident at once. The points of view he expressed were quite close to those he developed in the article intended for *IEEE Expert*:

> The theorem in my paper is correct, and its proof does not depend on full classical logical equivalence. Any notion of equivalence that includes any of many specific plausible equivalences, of which not (A and not B) = B or (not A and not B) is one example, also gives a collapse to two truth values.[55]

This passage used formulations that were present in the text intended for *IEEE Expert*. Only one expression was slightly modified. The logical equivalence referred to by Elkan is no longer characterized as "apparently reasonable" but as "plausible." It was thus a relatively inconsequential correction, if only because the message conveyed remained just as polysemic as before.

But Elkan's response was not limited to the issue of the correctness of his theorem. He again brought up the little problem of assessing the color of a collection of watermelons. According to the author, neither probability theory nor the notion of vagueness offered by fuzzy logic could be used to address this case. Elkan maintained that a third type of uncertainty was in play, linked to a notion of strength of evidence. This example showed, then, still according to Elkan, that fuzzy logic could not be used to reason about all types of uncertainty. The author fell back on an already-formulated conclusion, even if the argument was slightly modified, taking into account the new criticism of his eight contradictors. Once again, Elkan was maintaining an essentially stationary viewpoint.

[55] Ibid.

Finally, we see how the controversy tended to become stabilized in this forum in a very particular way. No consensus was reached as to the correctness of the proof. In contrast, the debates were immobilized in an opposition between two supposedly unique and well-identified viewpoints: Elkan's and that of the spokespersons for fuzzy logic.

This stabilization appears to be the fruit of convergent de-monstrative approaches. On the one hand, we see the emergence of a coalition of researchers in fuzzy logic ready to formulate a common critique analogous to the one Ruspini presented on the electronic forum, and prepared to reiterate the same objections in various other forums, thus helping to marginalize Elkan's positions. On the other hand, we see that Elkan's "mass" responses also tended to be anchored in a single set of formulations, those developed in a text intended for the journal *IEEE Expert*.

Turning back now to the site where the debates broke out in the first place, comp.ai.fuzzy, let us attempt to determine whether the tendency toward stabilization is confirmed. To this end, we can study a section of the FAQ created in May 1994 that proposed to draw up a balance sheet on the controversy.

Devices of Reiteration and Reference Contributing to a Stabilization of the Debates

From early 1994 on, the volume of communications devoted to the controversy on the electronic forum fell sharply, as we have seen. The contributions consisted mainly of compilations of messages on the topic that had been exchanged earlier. They responded to requests from actors who said that they had not followed the debates or had not retained copies of the communications and wanted to know the arguments that had been put forward on both sides. The regular formulation of such requests finally led the managers of the FAQ to post a section devoted to the topic in May 1994. They introduced their summary in the following terms:

Subject: [21] Elkan's "The Paradoxical Success of Fuzzy Logic" paper

The presentation of Elkan's AAAI-93 paper Charles Elkan, "The Paradoxical Success of Fuzzy Logic," in *Proceedings of the Eleventh National Conference on Artificial Intelligence*, 698–703, 1993, has generated much controversy. The fuzzy logic community claims that the paper is based on some common misunderstandings about fuzzy logic, but Elkan still maintains the correctness of his proof. (See, for instance, *AI Magazine* 15(1): 6–8, Spring 1994.)

We note in this passage how the FAQ, too, helped stabilize the controversy. It did not mention the considerable heterogeneity of viewpoints expressed earlier on the forum; instead, it presented the image of an opposition between an isolated author, Elkan in this instance, and a group united behind common objections, "the fuzzy logic community." All researchers in fuzzy logic, without any distinction whatsoever, were gathered together for the occasion under the same banner.

The representation of the controversy that had been offered in *AI Magazine* was faithfully reproduced. The repercussions of the pursuit of the debates in that journal could thus be observed (the reference in the FAQ to articles from this periodical in turn gave them additional weight).

The authors of the FAQ may well have been significantly influenced by the articles that had appeared in *AI Magazine*. For the similarities do not stop with this one aspect. The presentation of the controversy in the FAQ focused, as in *AI Magazine*, on the question of the correctness of Elkan's proof, leaving aside other issues that had come up in the course of the debates:

> Elkan proves that for a particular set of axiomatizations of fuzzy logic, fuzzy logic collapses to two-valued logic. The proof is correct in the sense that the conclusion follows from the premises. The disagreement concerns the relevance of the premises to fuzzy logic. At issue are the logical equivalence axioms. Elkan has shown that if you include any of several plausible equivalences, such as not(A and not B) $\Leftrightarrow$ (not A and not B) or B with the min, max, and 1- axioms of fuzzy logic, then fuzzy logic reduces to binary logic. The fuzzy logic community states that these logical equivalence axioms are not required in fuzzy logic, and that Elkan's proof requires the excluded middle law, a law that is commonly rejected in fuzzy logic. Fuzzy logic researchers must simply take care to avoid using any of these equivalences in their work.

This passage, too, contributed to the stabilization of the controversy. In this excerpt, the representation of the debates over Elkan's proof corresponds essentially to the viewpoints expressed in *AI Magazine*. The author or authors of the FAQ brought in a hybrid vision located somewhere between Elkan's and that of the spokespersons for fuzzy logic (while remaining closer to the latter).

The FAQ's reuse of the formulations provided by the researchers in fuzzy logic is especially noticeable in the presentation of the axiomatic system associated with fuzzy logic in Elkan's proof. This axiomatics was presented as general in scope, in Elkan's 1993 article; in the FAQ, it was characterized as specific to Elkan's proof. Similarly, the proof's "formal correctness" was acknowledged, but the implicit introduction of the principle of the excluded middle was once again denounced.

However, certain of Elkan's formulations were taken up again as such in the FAQ. For example, the form of logical equivalence so heavily criticized during the debates was characterized here as "plausible." Similarly, the idea that the users of fuzzy logic must "simply" be careful not to use certain logical equivalences converged with a viewpoint defended by Elkan.[56]

The FAQ thus had the effect of highlighting unanticipated (perhaps even forced) convergences between Elkan and his opponents, and, as a result, stabilizing the controversy even further. To be sure, allusions were made to the existence of differences other than the ones mentioned in the preceding passage. But the evocation of these disagreements in fact helped to reinforce the weight of the articles by the defenders of fuzzy logic, especially Ruspini's and Dubois and Prade's:

> It is difficult to do justice to the issues in so short a summary. Readers of this FAQ should not assume that this summary is the last word on this topic, but should read Elkan's paper and some of the other correspondence on this topic (some of which has appeared in the comp.ai.fuzzy newsgroup). Two responses to Elkan's paper, one by Enrique Ruspini and the other by Didier Dubois and Henri Prade, may be found as ftp.cs.cmu.edu:/user/ai/areas/fuzzy/doc/elkan/response.txt
>
> A final version of Elkan's paper, together with responses from members of the fuzzy logic community, will appear in an issue of *IEEE Expert* sometime in 1994. A paper by Dubois and Prade will be presented at AAAI-94.

In this passage we see how the FAQ, in referring to the writings of known representatives of fuzzy logic, constituted a device that helped reinforce the visibility of their arguments. But we can also measure the importance of the determination of prominent figures in fuzzy logic to express themselves in various context (journals, conferences, electronic forums and databases) in order to counter Elkan's de-monstrations with their own firm and stable positions. For example, the paper presented by Dubois and Prade at the AAAI'94 conference, cited in the excerpt above, was completely in line with the counteroffensive described previously; for this reason, we need not linger over its content. In contrast, the special issue of *IEEE Expert* devoted to the continuation of the debates cannot be situated so simply within this tendency to stabilize the controversy; thus it will be useful to clarify its tenor.[57]

Some of the authors of articles that had appeared in *AI Magazine* also contributed to *IEEE Expert*, presenting arguments comparable to those they had expressed earlier. However, other major figures in fuzzy logic, as well as several defenders of Elkan's theses, also joined in the debates.

---

[56] Let us recall, too, that Elkan presented the object of his de-monstration in an electronic message sent to me in the following terms: it was necessary to show "average users" of fuzzy logic, who were not necessarily "aware" of it, that certain logical equivalences could not be mobilized.

[57] *IEEE Expert*, August 1994.

By devoting an entire issue to the controversy, *IEEE Expert* could provide a forum where a large number of researchers could have their say.[58]

Even though co-optation could have played a role in the definition of the list of participants, the debates did not unfold in as "harmonious" a manner as in the other forums. In *AI Magazine*, only two antagonistic viewpoints emerged: Elkan contributed as an isolated author, and researchers spoke with a single voice in the name of a fuzzy logic "community." In *IEEE Expert*, other voices made themselves heard. Defenders of fuzzy logic and researchers supporting Elkan's theses, in some cases perhaps responding to the latter's encouragement to express their views, presented clearly distinct opinions in this journal.

Seventeen articles appeared in the special issue, with twenty-three signatories.[59] The interventions reopened the debates in several directions, although to a much lesser degree than on the electronic forum. Moreover, the researchers in fuzzy logic who had written articles for this issue were for the most part leading figures in the field. We need not examine the texts of these authors in detail, for we would uncover nothing really new in relation to the foregoing analyses. In contrast, it seems essential at this point to examine the outcome of the process we have been studying.

The shifting of the debates into journals starting in 1994 led by and large to a stabilization of the controversy. Overall, we witness both a decline in the number of different viewpoints expressed and a repetition of similar arguments in various forums. The controversy ends up running out of steam.

Even if the coordination among the viewpoints expressed was somewhat looser in *IEEE Expert*, we are still dealing essentially with a head-to-head confrontation between two limited sets of antagonistic positions. A sort of impasse seems to have been reached. On the whole, the positions of both camps were "known," the formulations well-worn; the adjustments belonged to the past. No consensus finally emerges. The only outcome was a group of extremely diverse representations of Elkan's proof(s), of fuzzy logic, and of a magma of de-monstrations produced over time. We are far removed from the stories of scientific controversies that end in the total victory of one side over the other.[60] For if a form of

[58] Including perhaps a sociologist of science, as noted above.

[59] See especially Hamid R. Berenji, "The Unique Strength of Fuzzy Logic Control," *IEEE Expert*, August 1994, p. 9; Enrique H. Ruspini, "On the Purportedly Paradoxical Nature of Fuzzy Logic," *IEEE Expert*, August 1994, pp. 32–33; Didier Dubois, Henri Prade, and Philippe Smets, "Partial Truth Is Not Uncertainty: Fuzzy Logic versus Possibilistic Logic," *IEEE Expert*, August 1994, pp. 15–19; L. A. Zadeh, "Why the Success of Fuzzy Logic Is Not Paradoxical," *IEEE Expert*, August 1994, pp. 43–46.

[60] Note that my remark is of course not intended to deny the existence of total victories of one side over the other in scientific controversies.

closure seems to have been established for a time in certain forums around two and only two competing positions,[61] such closure seems to have been suddenly called back into question when several researchers succeeded in taking the floor once again in the special issue of *IEEE Expert*, and very sporadically in later years.

As noted above, from 1994 to the time this book went to press, I was able to follow the traces of comments on Elkan's article and citations in comp.ai.fuzzy and in specialized journals. A few articles were also devoted to the continuation of the debates.[62] On the whole, however, the controversy's loss of steam was tangible. The arguments advanced were not really new ones. They did not trigger any new "flurry" of interventions. The representations of Elkan's theorem seem to have evolved very little after 1994. In particular, the section of the FAQ on comp.ai.fuzzy devoted to the controversy had undergone no modifications as of 2007. The outcome of the process, as described here, thus seems to have been quite "durable." The consequences of this affair on the activity of the actors' fields of research, reputations, and careers are difficult to pinpoint, but in any case they have been unspectacular. In particular, nothing indicates that the controversy had a determining impact on Elkan's career. In 2007 he was still at UC San Diego and had been promoted to the rank of full professor.

Regarding the case we have studied, we thus note that no "striking truth" managed to spring forth of its own accord and impose itself on everyone, to bring the debates to an end. In other words, no argument appeared capable of cutting the discussion short once and for all. We have in fact witnessed a shift away from such a model. In particular, with the use of heterogeneous de-monstrative practices, we have seen how a reduced number of firmly anchored antagonistic viewpoints emerged on the stage of the debates, with certain actors taking over as spokespersons for others.

In one sense, "Elkan's theorem" finally did achieve the status of certified knowledge. In fact, a certain number of actors finally agreed in a quite particular manner about "its correctness," and did so in a relatively lasting way. This agreement stemmed more from a form of loose coordination than from a convergence of views. If some of the protagonists agreed to recognize the "correctness" of "Elkan's theorem," this was only on the basis of divergent representations of what constituted that theorem and its correctness;[63] it was not because they finally managed to sublimate

---

[61] The possibility of reaching an analogous "settlement" of a mathematical controversy has been envisaged in Bloor, "Polyhedra" (1978).

[62] See again in particular the references indicated in note 41, above.

[63] For example, see again Ruspini's message on the electronic forum, where he evokes a trivial result (message 816, cited in chapter 3, n. 100); see also Elkan, "Paradoxical Suc-

the set of viewpoints expressed during the debates. Elkan did not construct a univocal text that satisfied everyone's objections.

At the end of the process studied, "the correctness of Elkan's theorem" appeared essentially as a *collective statement*[64] that each actor appropriated in a specific way. Around this statement, Elkan was able to find grounds of agreement, at least seemingly, for the protagonists attributed diverse and sometimes antagonistic meanings to the statement.

In any case, we certainly do not end up with a "universally" recognized statement. Indeed, in the final analysis, many participants in the debates considered that "that" theorem was not correct (once again lending variable meanings to this observation). Others took no stand on the issue. And, most notably, a still greater number of actors simply remained unaware of the existence of "that" theorem.

Such is the complex reality that a notion of the relative recognition of Elkan's theorem has to incorporate. This reality could not have been grasped without a fine-grained analysis of the details and intricacies of the de-monstrative dynamics, of the ways in which certain actors appeared on the scene of the debates or disappeared from it, and of the devices operating to stabilize the controversy. The task that remains is to recapitulate the overall results of this study and to see what conclusions may be drawn from the dynamics it has brought to light.

---

cess," August 1994, or the May 1994 FAQ of comp.ai.fuzzy, which put very different visions into play.

[64] See again Boureau, "Histoire restreinte des mentalités," and Alain Boureau, "L'adage *Vox Populi, Vox Dei* et l'invention de la nation anglaise (VIIIe-XIIe siècle)," *Annales ESC*, nos. 4–5 (1992): 1071–89: "I call a collective statement . . . a verbal (or iconic) fragment that creates around itself a certain convergence of languages and thoughts, through the play of a structural vagueness capable of seizing a still implicit thematics and accommodating the most varied projections and appropriations" (p. 1072). This notion appears better suited to account for the situation we have studied than the somewhat similar notion of boundary object. On this subject, see S. L. Star and J. R. Griesemer, "Institutional Ecology, 'Translations,' and Boundary Objects: Amateurs and Professionals in Berkeley's Museum of Vertebrate Zoology, 1907–1939," *Social Studies of Science* 19 (1989): 387–420.

CONCLUSION

A SOCIOLOGY OF THE PRACTICES OF DE-MONSTRATION

WHEN I BEGAN this study, I wondered whether it was possible to grasp, from a sociological perspective, the way a logical theorem was actually produced, and how it could achieve the status of certified knowledge. I tried to show it was possible by placing my study at the intersection between an analysis of technologies of proof and an analysis of forms of ostentation. As the great diversity of practices used in proving came to light, we were able to observe that the demonstration of a logical theorem required a vast set of heterogeneous resources. The most immediate contribution of this research is doubtless simply the fact that these resources have been observed and analyzed *in practice*.

The process studied turned out to put into play a chain of mediations, each important in itself and irreducible to any other. It is not possible to account fully for this dynamics without describing the nature, arrangement, and relative importance of the various elements in the chain, as I have sought to do. Thus I shall not attempt to *stylize* the case I confronted at this stage. Instead, we end up with a description of the unfolding of debates over a logical proof that a whole series of reductionist representations cannot capture.

DESTYLIZING THE UNFOLDING OF DEBATES IN LOGIC

First of all, the process studied cannot be reduced to a debate over "immaterial" ideas, or to an exchange of arguments expressed orally in some idealized locale. The debate we have been following bears little resemblance to the scenes in Plato's *Dialogues* or to the multi-voiced exchanges that Lakatos imagined in order to account for the controversy over Euler's theorem.

As I have sought to show, the activity of the participants in the debate over Charles Elkan's theorem involved writing much more than speaking. Several of the actors clearly perceived the importance of the material dimension of the exchanges, and managed their interventions accordingly; they also took into account the unequal distribution of re-sources among the participants. Factors such as these were fundamental in the economy of conviction.

Several elements thus distinguished the process I have described from a simple exchange of "immaterial" ideas. First, we can point to various competing material practices of writing and reading. Secondly, we have to take into account the resources needed to access the texts, the strategies of visibility deployed by some of the protagonists, the production of for-mulations differentiated according to the forums and the interactions in which the protagonists were involved. In addition, the effects of imperfect readings[1] on the part of the intended audience for these texts have been underlined, along with the effects of polysemic formulations. These latter were potential sources of an attenuation of the conflicts and important operators for rallying participants around a statement endowed with vari-able meanings. As a result, the exercise of logic appeared irreducible to a notion of "reasoning."

The process studied could not be reduced, either, to a simple confronta-tion between *individuals*. Group dynamics, especially in the formation of coalitions, the inscription of exchanges in conflicts over a long term and on a large scale, and the anticipation of the repercussions of the debates on group activities were essential elements. Moreover, the protagonists could not be viewed as homogeneous *subjects*. From the outset, they ap-peared to be endowed with extremely varied goals and resources.

Moreover, if I have already referred to the existence of an economy of conviction, I am not asserting that the debates unfolded *simply* in the mode of conviction. The fact that the controversy was stabilized around several viewpoints did not signify that all the protagonists had been con-vinced by the corresponding demonstrations. The evolution of the debates *also* hinged to a great extent (although, here again, *not exclusively*) on the fact that some participants managed to remain active and express their viewpoints while others did not, so that *in the end*, only a small number of positions emerged.

Rematerializing the debates also allowed us to escape from a different form of reductionism: social relativism. In fact, bringing to light the role

---

[1] I use this term to stress the fact that the readers had limited resources at their disposal (especially in terms of time and skills) when they sought to understand the proofs. Let us recall that in economics, the introduction of notions such as "imperfect money market" attempts to account for the limitation of monetary resources and for the consequences of this constraint on the dynamics of the economy (whereas in a perfect money market the availability of these resources would be total). One may estimate that, similarly, the limita-tions of the analyses produced by certain epistemologists stem from the fact that they implic-itly perceive the elaboration of knowledge in the context of a perfect reading market.

of written proofs made it possible to avoid reducing the unfolding of the controversy to a mere matter of social relations in the strict sense of the term. The constitution of statements was not purely *relative* to preexisting relational configurations among individuals or groups.[2] The texture of the debates and their unfolding via operations of writing and reading played a *determining* role in this dynamics.

Taking a stance opposed to social relativism made it possible to do justice to the fact that the smallest details of the proofs could intervene decisively in the evolution of the protagonists' positions. Nevertheless, I did not settle for a semiotic analysis of the texts grounded in the hypothesis of an ideal univocity of the symbolic inscriptions. I have shown that the texts did not automatically produce the effects that might have seemed to be programmed in them, particularly given the diversity of skills possessed by their readers.

In fact, the proofs could be compared to instruction manuals for technological devices. They partially determined their own interaction with their readers by delivering instructions for their own reading. Although useful, these instructions were insufficient to allow readers to replicate the precise modes of reading sought by the authors (such replication was in reality quite rare).

In the effort to account for such phenomena, Proust's analyses proved more helpful than some of the approaches developed in the semiotics of mathematics.[3] For what appeared in some people's eyes as immediately self-evident was perceived differently by others. Statements generally had to be abundantly accompanied, in particular by personalized reformulations, in order for a given meaning to be attributed to them, or a particular form of value or correctness. Only in exceptional cases did inscriptions suffice, unaided, to lead the reader from the beginning of a demonstration to the end in such a way as to produce a given reading.

That considerable difficulties arose in the endeavor to teach logic at the university level is finally not at all surprising. After assessing the diversity of the practices of logic and their often antagonistic character, we can understand why logic is *debated*. This fact might have appeared paradoxical if we had supposed that logic was a field based in practice on the production of immediate and universally shared instances of self-evidence.

---

[2] On the limits of social constructivism, see also Ian Hacking, *The Social Construction of What?* (Cambridge: Harvard University Press, 1999).

[3] It is conceivable, moreover, that the sciences of communication might attribute a leading role, on grounds other than that of the production of logical statements, to analyses that automatically take into account elements such as sources of "misunderstandings" about signs that may seem the most univocal, or devices for simultaneously diffusing different messages to distinct audiences.

FOLLOWING THE PRODUCTION
OF CERTIFIED KNOWLEDGE IN LOGIC

The analysis developed here has made it possible to stress the reductionist character of a number of representations of the unfolding of debates about a logical proof. It has also provided an alternative to some of the standard models for representing the way a logical theorem is produced and the way such a theorem can achieve the status of certified knowledge.

To the extent that the process studied put the activity of writing into play, it appeared neither "immaterial" nor simply localizable in the "minds" of one or several participants (which would have made any attempt at sociological investigation useless). On these grounds, I concur with Hutchins's conclusions on the necessity of *restoring* eyes and hands to those who manipulate formalisms.[4] Like other occupations that entail a considerable amount of writing, the exercise of logic was observable as it was taking place at workstations. It gave rise to man-machine interactions that could be observed on site. The study of exchanges on an electronic forum even made it possible to identify certain properties of a "distributed cognition" system (to borrow Hutchins's terminology), by underlining the way a group apprehended a particular statement.[5]

The results obtained show, moreover, that the work of logicians can be the object of fruitful investigations in cognitive science. One of the research paths that stands out at the end of this study imposes a double displacement with respect to a certain number of cognitive approaches to logical manipulations. The first consists in observing the manipulation of formalisms "in the field," and not simply in laboratories, with their experimental arrangements. The second lies in studying the specific capabilities of logicians themselves, as opposed to "ordinary subjects," to manipulate symbols. Research along these paths would not simply entail analyzing the way logic "in general" is apprehended and utilized from the

---

[4] It is also appropriate to reattribute screens and keyboards to computers. Such an undertaking could be seen as an extension of Hutchins's computational model; see *Cognition in the Wild*, pp. 353–74. Of course, these conclusions apply in very large measure to the analysis of various forms of the work known as "intellectual."

[5] I am referring to the system formed by the participants in the debates and their computer devices. This system could be endowed with its own cognitive capacities, distinct from those of its components. It should be noted that the system I observed was quite different from the ones Hutchins examined, in that its operation was marked by *competing* efforts at orchestration. Far from being "collaborators," the participants were, to some degree, rivals. For this reason, it would have been useless to try to describe the system's operation simply in terms of coordination. The examination of Elkan's case shows that the analysis of phenomena of competition constitutes a promising axis of research for the study of distributed cognition.

starting point of a representation of logic that has been *fixed in advance* (for example, a representation based on syllogisms). The goal would be to seek a better grasp of what "professionals" in logic actually do, without taking for granted any idealizing representations that attempt to go directly—perhaps too directly—to what is "essential."

It would be particularly valuable to be able to identify better the skills, the routine tasks, and the durable or transitory phenomena of apprenticeship that come into play in the exercise of logic. For example, one could seek to spell out the way logicians learn *to see how to make* symbols *appear,* along with the nature of the repetitive practices that are potentially involved in these processes. It might also be fruitful to analyze systematically the whole set of *residues*[6] from this exercise, and also to try to look more thoroughly into the ways self-evidence is experienced, without neglecting all its emotional dimensions.[7]

It would of course be important to develop models that bring to light the collective (and not merely individual) dynamics that come into play. For if the specific process through which Elkan's theorem was produced (even before it attained the status of certified knowledge) did not appear localizable in the mind(s) of one or several actors, this process did not turn out to be reducible, either, to a purely "individual" dynamics. On the contrary, we have seen that from the outset it involved a certain number of interactions and interventions on the part of groups of actors. These groups played a determining role both in the process of revising drafts of Elkan's proof and in the dynamics leading to publication.

For this reason, the decision to publish Elkan's text was difficult to interpret. For the analyst as for the actors, it could not be equated with a guarantee of universal consensus on the question of the proof's correctness. It referred back to the traversal of a series of varied trials, and especially to convergences between Elkan's viewpoints and objectives on the one hand and those of reviewers and program committee members on the other. Such an outcome was not *readable* for anyone who had not followed the various stages of the corresponding process very closely.

Thus we have not been dealing with a theorem grasped in a single way, uniformly accepted or rejected by its entire readership. By following the different stages of the process, grasping the protagonists' diverse translations in their specificity without representing them as equivalent, I have been able to bring to light the formation of a plurality of viewpoints on the content of Elkan's theorem and its proof, as well as on its correctness and its value.

---

[6] This term refers in particular to the written productions of the actors, to the constitution of skills, and to the construction of more or less durable systems of representation. On this subject, see Hutchins, *Cognition in the Wild*, pp. 263–85.

[7] Ibid.

For, while a certain number of participants in the debates did end up acknowledging the "correctness" of this theorem, one cannot speak of consensus on the subject. The apparent agreement depended in fact on the production of a "magma"[8] of de-monstrations presented as faithful substitutes for Elkan's proof, that is, on the coordination of radically opposed viewpoints as to the nature of the theorem, its meaning, and its scope. By grasping "the correctness of Elkan's theorem" as a collective statement, I have been able to account for an original form of rallying on the part of actors around a statement, as these actors appropriated it for themselves while endowing it with varied meanings. It is in this sense that I have finally been able to attribute meaning to notions such as "relative recognition" and "certified knowledge," and to use them to characterize the status acquired by Elkan's theorem in the eyes of some of the protagonists in the debates.

I have thus described a process of production of certified knowledge without having had to "explain" its unfolding by evoking in a central way the *ideal* correctness, or absence of correctness, of particular "arguments." This position was all the easier to adopt in that the researchers I encountered did not limit their questioning to the correctness of proofs. They introduced viewpoints that were quite tangential with respect to that issue, and they debated a wide variety of elements.

Correlatively, I have not had to denounce the irrationality of certain actors and to evoke phenomena such as "diffusion of ideas" or "resistance" in that connection. In other words, I have been able to keep my distance with respect to the interplay of the protagonists, who often apprehended the positions expressed by their adversaries from a relativist perspective, even as their belief in their own viewpoints remained "rock solid." Their observations of the wide disparity of convictions among their interlocutors generally did not affect their own attitudes. In the face of the entire set of statements, only the culturalists (as opposed to the universalists) adopted stable postures, those of reflexive relativists.

In the final analysis, no structure of a universally recognized ideal logical world and no preestablished and stable social configuration as such emerged that might have made it possible to determine the outcome of the controversy. I witnessed a confrontation between diverse conceptions of what logic "in general," fuzzy logic "in general," and a logical proof should be, in the context of an ongoing and general negotiation of ways of interacting (including with texts). It was during this process that some

---

[8] The term "magma" is used to designate a gradual accumulation of de-monstrations capable of creating confusion about the object of a statement, and, in consequence, capable of leading to agreement on the latter. The meaning I attribute to this notion is comparable to the use made of it in Greek rhetoric. On this topic, see Barbara Cassin, *L'effet sophistique* (Paris: Gallimard, 1995).

logical practices turned up on center stage and that stabilized representations of logic and of the correctness of certain statements imposed themselves. *Simultaneously*, coalitions and modes of interactions emerged,[9] underwent transformations, and were stabilized.

The settling of the controversy, then, ultimately helped produce and modify both relational systems and relatively fixed representations of an ideal logical world. This result depended on the intervention of an operator that is as essential as it is little studied in the social sciences, one that has the particular ability to constitute and transform associations both among statements and among individuals or groups of individuals, the operator I have called de-monstration.

OBSERVING THE WORK OF DE-MONSTRATION

One of the most important factors behind the evolution of the Elkan controversy lay in the participants' *de-monstrative* activity. Let us recall that a special meaning is attributed to that term here; it seemed appropriate to attach to the notion of demonstration a set of activities that are not generally associated with it. Conversely, it was necessary to find a vocabulary for characterizing different exercises packed together under this same term. This is one of the reasons for the introduction of the word "de-monstration": I sought to avoid any confusion between the two terms and to stress the specific use I was making of this notion.

Ultimately I was able to distinguish several types of de-monstrative practices, and, correlatively, to specify several declensions of the term "de-monstration," such as *cause to appear, bring to the fore, bring to light, make visible, put forward, put on display, and transformalize*. However, I was also able to highlight the use of a set of heterogeneous registers (for example, the statement of theses pertaining to the history of science and the history of civilizations, or the formulation of political denunciations) that the reader might not have spontaneously associated with a demonstration or counterdemonstration of a logical theorem. In fact, the list of registers deemed to stem from the exercise of logic was redefined, with important variations, by every participant.

In the course of the narrative, I have sought to draw up a tableau of logical activity that assigns a place to the set of de-monstrative practices I observed and that recaptures their arrangement. Beyond their diversity, these practices had in common the fact that they involved a "monstrative" undertaking: One important aspect of these practices was the act of show-

[9] For example, see the analysis of the meeting between Elkan and Peter Turner in chapter 6, above.

ing. Thus I had to reinterrogate a dichotomy, highly sedimented in the philosophy of science, between showing and demonstrating. The term de-monstration was introduced to describe series of moments in which the authors judged that they could be content to *show*, series whose main declared purpose was generally to prove something. We were then confronted with sequences of de-monstrative *steps*. Under these conditions, it appeared possible to define one type of "formalist project"—which was partially implemented in the actors' de-monstrations—as an attempt to *make everything visible*.

Generally speaking, the de-monstrative practices stemmed from routine exercises. Combinations of monstrative forms were used repetitively, like stylistic figures in rhetoric.[10] *Making self-evident* by writing that such-and-such is the case constituted one particular monstrative form.

The work of de-monstration was marked in addition by the importance of rewriting proofs and of the corresponding adjustment procedures. These were paired with acts of visual control of symbolic manipulations. The readings of the de-monstrations were carried out with the help of signaling systems present in the proofs themselves. These systems defined only in part the ways in which readers were supposed to interact with the texts. Like instruction manuals for technological devices, they were useful but not entirely sufficient for decoding the proofs.

To be able to read the texts the way their authors wanted them read, and to be able to grasp the necessity for each de-monstrative step, a large number of tacit skills were usually required. By this token, the de-monstrative steps possessed the characteristics of a quasi-object. They could (and did) constitute the object of sociologizing or objectivizing ways of seeing on the part of the protagonists. We thus understand why the question of the necessity of a proof in mathematics and in logic could be treated in a purely objectivizing or sociologizing way by philosophers and sociologists of knowledge. A *nonfrontal* and *nonstylized* approach to this question constitutes an essential stage if we are to move beyond the debates associated with confrontations between such ways of seeing.[11]

[10] It is likely, moreover, that the work of interrogating representations of rhetoric in the social sciences is as difficult as the work to be done for logic. In particular, the Greek philosophers seem not to have done justice to competing forms of rationality when they presented the Sophists as enemies of reason. On this subject, see Cassin, *L'effet sophistique*.

[11] The same thing holds true for the question of what lies behind the "beauty" of a musical work. On this subject, see Hennion, *La passion musicale*. The comparison is instructive to the extent that reading a logical proof seems to be somewhat analogous to "interpreting" a musical work. In both cases, a considerable number of mediations are put into play. During my research, I observed that the analysis of the mediations used in logical activity could not be limited to the examination of reading practices but had to include writing practices as well. Thus it seems plausible that, similarly, a sociology of musical mediation cannot do without a study of the situated practices of musical composition and the material procedures

The de-monstrative practices that we have analyzed were based in large part on activities of reading and writing that are also characteristic of other forms of "theoretical work." Thus it appeared clarifying to speak of *theoretical practices* in this connection. This expression makes it possible to emphasize that the form of theoretical activity with which we are concerned is not abstract *in itself*.

To the extent that the practices in question are themselves material,[12] it is neither necessary nor relevant to reduce them to *other* material manipulations, by presenting them, for example, as a "reflection" of these latter.[13] To proceed in this way would amount to adopting an approach analogous to that of the researchers I met, with the risk of obscuring the reality of the practices in question. In fact, we are dealing here with the work of translation carried out by the actors *themselves*, and this work represents a crucial dimension of their activity. The protagonists in the Elkan controversy often defined their objects by creating associations between the "formal" manipulations introduced and "natural" operations. In other words, the symbolic procedures put forward were involved *from the outset* in specific practices and translations. For the analyst to present them as equivalent to other manipulations would not have done justice to the latter.

The specificity of my object of inquiry imposed a quite specific methodology. It required me to *follow* the de-monstrations and what they *caused to appear*. Under these conditions, I myself in turn had to *make* my readers *see*. By this token, the status that I attribute to my own de-monstration is no different from the one I attribute to the de-monstrations of the participants I met. Like them, I had to constitute *monstrative* sequences and use comparable methods to de-monstrate—although not methods relying on the use of symbolic manipulations. Thus I do not claim to have achieved a result that is different in nature from the ones I observed. In particular, it is entirely conceivable that this work may be subject to the test of contradictory ways of seeing, and that I may find myself in the position of having to accompany it by becoming its spokesperson, by producing new de-monstrations, orally or in writing.

Interrogating the modalities of de-monstrative activity in the so-called deductive sciences seems finally just as important and fruitful as focusing—as many sociologists and historians of science have done—on the modalities of replicating an empirical proof in the so-called experimental

---

used in writing music (in cases where such a description is valid, of course). This dimension must not be neglected in particular when it is a matter of drawing up the list of operators of stabilization of taste, a list that is doubtless as long as the list of operators of stabilization in debates in the field of logic.

[12] I use the term "material" only to the extent that I believe that some might characterize the reality in question—inaccurately—as "immaterial."

[13] See above, pp. 25–26.

sciences. The stabilization of a logical statement has in fact appeared to involve operators as heterogeneous (and perhaps as unexpected) as those that come into play in the sedimentation of experimental facts.

Even as it helps broaden the perspectives of sociological and historical inquiry into the sciences,[14] this research also raises questions about ways of describing the way scientific phenomena are consolidated. Many studies of scientific controversies present forms of closure through the total victory of one group over another, or several others: the entire set of protagonists ends up transcending earlier divergences and adopting univocal statements, as transportable as immutable solids.

The present study helps bring a different sort of phenomenon to light, one that resides essentially in an apparent convergence of distinct and largely antagonistic viewpoints, and in a phenomenon of co-ordination around a collective statement "accommodating the most diverse projections and appropriations" (to borrow Alain Boureau's terms once again).[15]

The statement "Elkan's theorem is correct" was by no means adopted by the *entire* set of actors. In the end, this phrase did not subsume their respective viewpoints. It was neither apprehended in a single fashion nor transported as an unalterable object. And yet, owing in particular to operations of adjustment and polysemic reformulation, and in particular without going through a simple process of *interessement*,[16] it managed to become the rallying point for a certain number of participants in the debates.

LOGIC AND SOCIOLOGY

The results of this research take us far from epistemological models that reduce the constitution of knowledge in mathematics and logic to the accumulation of theorems endowed with unique meanings, theorems ob-

[14] In particular, there is no reason for the sociology of science to limit its scope mainly to the experimental sciences.

[15] I should note that sociological investigations I have carried out into other dynamics of production of logical statements have allowed me to observe that the emergence of a collective statement around "Elkan's theorem" is not an isolated case. This observation may well represent an important direction for further research, especially inasmuch as the type of phenomenon brought to light has very little in common with the representations associated canonically with the nature of logical theorems. This being the case, I do not claim, of course, that logic is any less capable than other sciences of establishing perceptibly "harder" facts that can be mobilized widely through the mediation of potentially more substantial and more complex devices.

[16] Michel Callon defines the notion of *interessement* as a series of processes by which some actors seek to lock other actors into roles that are proposed for them. See in particular Callon, "Elements of Sociology of Translation."

tained ideally on the basis of "methods" that can be grasped schematically, with the "social" playing only a peripheral role in the dynamics.[17]
In particular, we are far removed from an a-sociological representation of
the ways logical knowledge is produced, based on faith in the autonomy
of symbolic demonstrations in logic. Starting from the latter sort of representation would have made it impossible to grasp the ways in which viewpoints become less numerous and the controversy becomes stabilized. To
do that, it was imperative to analyze in particular the effects of the polysemy of the proofs, perceive the uses of implicit formulations and reference in the face of the participants' imperfect readings, study the way an
author produced evolving tailor-made and mass-oriented responses while
taking into account the divisions among his detractors, pinpoint the nature and stakes of the formation of coalitions, and also evaluate the respective roles played by the texts researchers produced and the activities
undertaken to accompany those texts.

The approach I adopted cannot be characterized as an attempt to discover and denounce. The activity for which I was seeking to account was
not "hidden" by the protagonists. My description does not "unveil" elements that the actors would have preferred to conceal. On the contrary,
the variable lists of heterogeneous registers that they drew up to define
the legitimate exercise of logic constituted, for them, a permanent object
of *debate*.

It is quite conceivable that the results of this study may appear surprising to readers who have never been directly involved in research in logic
or mathematics. If they perceive this activity through the prism of the
training they may have received in these fields, or through the schemas
proposed by certain epistemologists (in particular those who have never
practiced research in mathematics or logic), such a reaction is even probable. But the fact that certain data may appear dissimulated from the viewpoint of some epistemologists who do not *take* them as an object does
not mean that they are dissimulated from the viewpoint of the researchers
themselves.

Moreover, actors in logic are often the first to assert that their work and
their objects are presented in a caricatural way in many epistemological
studies.[18] They themselves generally develop competing representations of

---

[17] Let us note that, given the rudimentary notions of the "social" put into play in these
models, their authors *rightly* credit it with a peripheral role.

[18] These denunciations are made orally much more often than in writing. Presumably,
this is largely because their authors do not consider these "misunderstandings" harmful to
their own activity, and thus deem it unnecessary to devote time to trying to dispel them.
However, see again Largeault, *La logique*, pp. 122–25; and Rota, "Pernicious Influence,"
p. 168.

their own activity in the context of their research and the accompanying debates.[19] This state of affairs certainly has to do with the resonance of a strong division of labor in science and with the weak diffusion of contemporary knowledge in logic, but also with the weak visibility of research practices in logic, especially among philosophers of science and researchers in the human and social sciences.

By presenting the results of my study, I hope to be making more visible a certain number of human practices[20] that are largely overlooked by the social sciences. To the extent that I have attempted to give visibility and meaning to a certain number of practices, but also to "modes of behavior," both permitted and restricted, to "forms of organization," to "stakes" advanced, to individual and collective "strategies" on display, to various forms of "professional investment," and to "modes of life" linked to a specific activity, and to the extent that, on the basis of the arrangement of all these elements, I have tried to reconstitute a particular world, characterized by certain irreversible phenomena as well as by constraints and by determined possibilities for action, my work may be viewed as sociological.

Nevertheless, my sociological project may be perceived at another level, one closer to the Durkheimian questions addressed in the introduction. By taking a direct interest in a dynamics of production of certified knowledge, I have tried to show how chains of credit are constituted. At a more fundamental level having to do with the formation of the social link,[21] I have sought to understand why actors *associated themselves* in this activity.

This sociology was intended to be respectful of the world constituted by its actors. That is why I did not reduce it to *human* relations. To avoid doing that, I had to recapture the material aspects of the work carried out, particularly the importance of writing in relation to oral work, countering the image of logical activity offered, for example, in Plato's *Dialogues*.

However, by helping to offer sociology a "new" object for empirical investigation, this study contributes elements for reflection about what is often presented as one of sociology's methodological objects. In fact, logic is often perceived in the social sciences as an instrument that can guaran-

---

[19] Moreover, certain of these representations have a fully epistemological vocation.

[20] With all the consolidations or transformations (difficult to pinpoint and to predict) that this undertaking, if it converges with others, may potentially imply in the long run.

[21] However, other types of links were involved at the same time, such as links between statements.

tee the coherence of analyses,[22] reveal their structure,[23] and make it possible to judge the rationality of the behaviors observed. Apart from a certain number of exceptions,[24] the representations of the objects of logic that one encounters often date back to Greek antiquity or to the end of the nineteenth century, and are most often not dynamic but stabilized. In the context of these representations, comparable to the logicist approaches that were denounced and then largely abandoned by logicians at the beginning of the twentieth century, logic is structured by the existence of a small set of immutable and ideal "principles" or logical "rules." Generally speaking, these rules are readily evoked and rarely problematized. Frequent reference is made to syllogisms, which are perceived as stable objects within the discipline.[25]

The practices I observed are quite different from the concepts and uses evoked above. The production of researchers in logic cannot be identified with the production of immediately self-evident rules that would impose themselves universally, nor can it be reduced to the production of tools of coherence. It does not consist simply in asserting that logic "in general" rests on the univocal meaning and universally recognized correctness of syllogisms, and that the mobilization of symbols favors consensus and is synonymous with immediate self-evidence. The fact that some sociologists can adopt representations of this sort underlines above all the current lack of visibility of contemporary logical activity in the social sciences. This shadowy zone is at least as important as the one that involves the operations and products of other fields of contemporary scientific research (physics and mathematics in particular).

Bringing to light the considerable gap between such representations and reality does not imply, of course, that one must give way to "skepticism." Moreover, the conclusion that this study consists in showing up the "inconsistency" of logic "in general" would derive from a misunderstanding of the worst order. If, by bringing to light the highly creative dimension of logical activity, my work succeeds only in stimulating or reviving curiosity

---

[22] A radical critique of any attempt to reduce the exercise of sociology to certain formal translations and manipulations can be found in Jean-Claude Passeron, Le raisonnement sociologique: L'espace non poppérien du raisonnement naturel (Paris: Nathan, 1991), pp. 154–60.

[23] This idea may be at the origin of enlightening analyses of the production of the social sciences, moreover; see especially J.-M. Berthelot, Les vertus de l'incertitude: Le travail de l'analyse dans les sciences sociales (Paris: Presses Universitaires de France, 1996).

[24] For just one example, see Pierre Naville, Sociologie et logique: Esquisse d'une théorie des relations (Paris: Presses Universitaires de France, 1982).

[25] See Claude Rosental, "Quelle logique pour quelle rationalité? Représentations et usages de la logique en sciences sociales," in L'argumentation: Preuve et persuasion, ed. Michel de Fornel and Jean-Claude Passeron, Enquête, vol. 2 (Paris: Éditions de l'École des Hautes Études en Sciences Sociales, Mar. 2002), pp. 69–92.

about contemporary research in this field and leads some researchers in the social sciences to wonder about the appropriateness of associating the search for consistency in their own writings, the evaluation of the rationality of the actors,[26] or even sociological work properly so-called, with the mobilization of some rules or syllogisms perceived as central or immutable in logic, it will already have been of some use.

For of course I have by no means accounted for the entire set of contemporary practices in logic; indeed, that was not the goal of this project. My aim was above all, more modestly, to advance knowledge of some specific human practices, and to bring to light the reductionist character of a certain number of models having to do with the constitution of knowledge in logic. After seeing that the production of certified knowledge in logic can constitute an object of investigation and sociological analysis, we can now contemplate the vast field of research that lies ahead.

If, in the context of this study, no logical concept or practice appeared as consensual or central (including "respect" for the principle of the excluded middle), what can be said about contemporary logical activity in general? Is the present result valid, as Jean Largeault suggests when he asserts that it is not possible to determine "central" elements in logic, for this field of research as a whole? A great deal of investigatory work would have to be done before this question could be answered. One would have to be able to determine on a broader scale what the work of contemporary logicians actually comprises, and even simply who these logicians are. In other words, one would need to establish a cartography of contemporary logical activity.

Inasmuch as no logical statement was an object of consensus in the context of the process studied, and inasmuch as considerations about cultures and the origin of the operation of technological devices were mobilized in order to question or defend the correctness of a theorem, it is appropriate to envisage calling into question our representations of scientific work as a whole. It seems particularly important to question the relevance of an architectonic vision of science that would be based on the universal recognition of some immediately self-evident logical statements, such as the principle of the excluded middle.[27] My research has not provided a full picture of what might be substituted for such a vision. But several results seem to indicate that the situation needs to be dedramatized, and that the place of logic in relation to other scientific disciplines

---

[26] The adoption of representations of logic as specific as the ones I have underlined is in fact quite consequential for the modes of evaluation of rationality in the history of the social sciences. The adoption of new representations of logic as such may lead to revising certain accusations of irrationality (see ibid.).

[27] Let us recall that Husserl presented this principle as one of the major ideal foundations of logic and science.

and to the history of civilizations and their future needs to be reconsidered. Let us recall, first of all, that the production of certified knowledge in logic has turned out to be capable of putting into play objects developed in places other than centers of calculation.[28] We have also seen that attributing the operation of certain electronic devices to a logical theory was not self-evident and produced no consensus; we have seen that symbolic manipulations could be endowed with effects that were limited if not null, and that logical statements could be the object of a splintering of strong but contradictory certainties.

History is full of cases in which we see logic—logic "in general" but also syllogisms—put on stage in a dramatic way.[29] From this standpoint, the destiny of science, or even of civilizations, is at stake; by making it possible to avoid erring discourses, logic "in general" constitutes one of the chief pillars for the maintenance of civil and world peace.

While my study has not allowed us to verify such grandiose assertions, I have nonetheless perceived some effects proper to symbolic manipulations, in the context of the interactions of readers with texts. For example, we have observed the reinforcement of overt viewpoints regarding a proof in connection with the production of new de-monstrative texts.

In addition, we have not witnessed any situations in which reference to Elkan's theorem would have made it possible to reinforce a decision having harmful consequences for the pursuit of research in fuzzy logic. But we have observed how overt anticipation of the negative impact of Elkan's theorem on the funding of projects in fuzzy logic was correlated with the production of counter-de-monstrative texts.

Moreover, we have noted that the use of symbols in the writing of a proof implied specific modes of reading, distinct from those induced by formulations in prose. Singular acts of visual control were made possible as soon as an elaborate system of signaling had been mastered, even though recourse to this tool did not suffice as such to make a proof's correctness or lack of correctness immediately clear, nor did it suffice to create a consensus around any particular way of doing logic.

It must be noted that my field of inquiry was distinctive in that it put into play a *production* of certified logical knowledge, and not simply the *implementation* of formalisms. For this reason, I was not able to bring to light the capacity of a formalism to constitute "freeze-dried" objects that could be transported into centers of calculation in such a way as to make a multiplication of the effects of a simple symbolic manipulation observable. In contrast, this displacement of the gaze allowed us to grasp an

---

[28] See again Latour, *Science in Action*.

[29] See again especially Hilbert, "Axiomatisches Denken"; Boll and Reinhart, *Histoire de la logique*; Bergmann, Moor, and Nelson, *The Logic Book*.

essential dimension of logical activity, the work of de-monstration. By studying a universe of practices that up to now had not been addressed by studies in the sociology of the experimental sciences, I went from a problematics focused on "formalism," on operations of *putting into form*, and on modes of coordination, to questions related to the details and intricacies of a de-monstrative activity. In other words, the particular world that I have described could be characterized not in terms of coordination but rather on the basis of the de-monstrative activities undertaken.

Moreover, the competitive character of the development and use of formalisms, as we have seen, seriously complicates the overall picture of the ways knowledge is put into place. The possibilities of coordination and multiplication of action through harmonious cascades of re-formalization turn out to be limited.[30] To discover what holds together chains of translation implying successive operations of re-formalization, one needs to be in a position to understand why certain formalisms rather than others are used for a given task. In other words, it is essential to take into account the dynamics of competition *between* formalisms (for example between fuzzy logic and "classical" logic for the development and presentation of controllers), even if these are off-center with respect to a given site of formalization. The analysis of de-monstrative activity constitutes a tool for grasping such dynamics.

In the context of the present study, this type of analysis has in any case made it possible to bring to light confrontations between antagonistic logical practices (characterized in particular as "formalist" and "logicist"), which various schools in the philosophy of logic have attempted, each in its own way, to define and constitute in part, and to which the protagonists in the Elkan controversy sometimes referred. Through normative displays of various ways of doing logic, we have encountered more or less remote and amalgamated translations of competing philosophies of logic. The participants were not all strictly "formalists," "logicists," or even "idealists," "realists," or "relativists." We have had the opportunity to grasp the diversity of means for building logical self-evidence envisioned by the actors. Having seen the considerable number of resources to be gathered in order to constitute the "blinding clarity" of immediate self-evidence, we can appreciate the meticulousness, subtlety, and quasi-heroic hope certain philosophers have had to manifest in order to establish, apprehend, and reforge in part very elaborate procedures of symbolic writing, intended in particular to avoid potential errors and erring ways in speech and writing.

Bringing to light the vast quantity of resources to be collected in order to shape the weaving of self-evidence also allows us to make sense of the

[30] See Latour, *Science in Action*.

denunciations leveled by logicians against certain philosophers who are said to fail to comprehend the nature of research activity in logic. It also provides tools for conceptualizing the proliferation of philosophies of logic. These latter, even as they themselves introduce new resources, account in their various ways for part of a vast set of mediations to be reckoned with, if not absorbed.

By allowing a positive grasp of the actors' ways of perceiving their own practices, my analysis thus converges to some extent with the representations of the researchers I encountered. From a methodological standpoint, this allows me to respond to one of the principal requirements (but also one of the principal difficulties) of ethnomethodology.

Toward a Social and Material History of Forms of De-monstration

By taking de-monstrative practices as my object, I was able to avoid reducing my questioning of "logical formalisms" to an interrogation exclusively focused on "formalization." The development of a social and material history of forms of de-monstration would give full meaning to this displacement, as well as to the methodological tools and descriptive language developed here. Such developments would entail retracing the situated historical evolution of modes of de-monstration in the so-called deductive sciences, without failing to take into account the material conditions of the activity of proving and the forms of association that this activity puts into play.

My study brings to light the deployment of forms of de-monstration whose emergence cannot help but transform the nature of logical activity, in certain spaces. Such is the case in particular for the de-monstrations involved in exchanges on an electronic forum or in computer demos.

We have seen how the mediation of an electronic forum authorized recourse to specific writing practices for debating proofs. In particular, it allowed passages from contested de-monstrations to be commonly inserted into counter-de-monstrations. The mediation of an electronic forum also offered certain readers the possibility of putting forward their own proofs or counterproofs, something they would not have been able to do if they had had to face the constraints imposed by journals.

In addition, the existence of such a forum constituted a supplementary means whereby the participants could *adjust* their positions and formulate de-monstrations adapted to their various interlocutors, in a Goffmanian context of nonunity in the presentation of self and work. Moreover, it led to the constitution of a tribune for antagonisms and it shaped the latter accordingly. It allowed confrontations between actors who would probably not have found themselves otherwise in a situation of (sometimes apparent) dialogue.

Recourse to the electronic forum also led to a profound modification in the pace of exchange and in the strategies adopted to gain the upper hand in the debates or at least to remain in the game. The phase during which the controversy unfolded in the pages of professional journals offered a first reference point that clearly underlined this phenomenon. But the history of science, too, supplies a large number of elements for comparison. These elements make apparent the transformation such a forum can induce in the nature of what a scientific "controversy" may be.

The debates studied took place essentially during a single year, rather than spanning a generation or more. For the protagonists, it was not a matter of drafting articles over a period of several years while going on to pursue other research interests and taking care to constitute "parachutes" and to be in a position to recycle their experimental equipment.[31] We cannot help but note, in comparison, the relatively *open* character of an electronic forum, which makes it easy for many to join in the debates and remain players; we also note the relatively small quantity of resources that such acts entail.[32]

The fact that the debates I was studying unfolded in part in the context of an electronic forum unquestionably made my task easier. Still, we must be careful not to conclude from this that the same type of study would have been impossible without the existence of such a forum. For several years, I pursued the inquiry on quite diverse terrains. In particular, I attended research seminars leading to the formulation of new logical statements. The observations made in that context offered elements that are just as relevant for the treatment of my problematics.

The existence of debates on an electronic forum thus did not constitute so much a necessary condition as an effective tool for analysis. For example, the preservation of traces of the exchanges in electronic archives made

[31] The history of the experimental sciences makes it possible to apply a dual principle of variation, inasmuch as the involvement of sometimes "heavy" technological devices in order to do experiments induces its own dynamics, something not found in the case studied here.

[32] In this connection, we must not rule out the possibility that the spread of the Internet or other electronic networks may lead to a transformation in the nature of scientific work that is at least as important as the emergence of printing; see Eisenstein, *Printing Press*. If the approach I have developed were extended to help study this evolution, it would be important not to concentrate exclusively on a *sociology of the uses* of electronic networks, that is, on a study immersed in the uses of this medium alone (something that seems to have constituted a "reflex" perceptible in several research projects on the topic). In contrast, it would seem appropriate to bring to bear on this subject a *sociology of mediations* that is attentive to the contexts in which these networks are preferred (or not) to other media (or used in combination with them) for the execution of specific tasks (for example, the "discussion" of a theorem).

it possible to reconstitute the unfolding of the controversy easily. In addition, the diversity of the participants in the debates in this context made the electronic forum an excellent place to observe a set of logical practices that I had found on other sites.[33] These practices could be observed in a single place, where they were, exceptionally, *gathered together*. Without the existence of this forum, to describe them all it would have been necessary to multiply the accounts of investigations.

Following a scientific conference firsthand also proved to be very fruitful. Indeed, I was able to observe actors from a worldwide network operating in a single location. One of the principal obstacles encountered by ethnographies of science has lain in the difficulty of going beyond local laboratory studies to apprehend associations in very widespread networks that are hard to explore extensively. From this standpoint, scientific conferences constitute privileged sites that warrant further exploration by sociologists and historians of science. The intensity of the interactions that take place at such meetings and the importance of the stakes for the participants only reinforce the value of the undertaking.

From the standpoint of the development of a social and material history of forms of de-monstration, the observation of the unfolding of a scientific conference like the one I attended is also very valuable. It allows us to pinpoint situated de-monstrative strategies that are tied to the overall objectives of the conference organizers but are also characteristic of the various participants. The procedures for selecting papers and the role played by the actors' expectations regarding the modalities of this process as they draft their texts for submission are also elements worth observing and analyzing.

In addition, the spread of technological demos, which constituted important forms of de-monstration during FUZCOL, represents an object of investigation in its own right for the treatment of this problematics.[34] Indeed, I have been able to observe frequent recourse to demos in other contexts.[35] To the extent that these forms of de-monstration were directed toward diverse audiences, sometimes consisting primarily of industrial-

---

[33] It is important to stress, moreover, that I observed not simply the practices of fuzzy logic here but various practices of logic. Only a certain number of the participants in the Elkan debates were actually researchers in fuzzy logic. It should also be noted that the type of debate I was studying was not unique, nor was it specific to the development of a particular form of logic (i.e., fuzzy logic). Other logical formalisms were the object of comparable controversies on several levels.

[34] For more results in the sociology of demos, see again Rosental, *Les capitalistes de la science*; Rosental, "Making Science and Technology Results Public"; Rosental, "Fuzzyfying the World"; Rosental, "De la démo-cratie en Amérique"; Rosental, "Histoire de la logique floue."

[35] For example, I was able to observe the use of computer demos during thesis defenses in logic.

ists, in order to solicit funding, they constituted an essential link between science, technology, and society.[36]

However, and in conclusion, it is probable that the development of a social and material history of forms of de-monstration will require investigations of activities usually viewed as nonscientific. It may prove fruitful to explore connections that may not be celebrated at a given moment between certain modes of de-monstration perceived as stemming from scientific work and practices judged extrascientific. For example, one might raise questions about the nature and the evolution of potential links between "demos" and the demonstrations made by salespersons. This type of exploration could well constitute an essential step in the effort to circumscribe, in turn, the peculiar activity we call reason.

[36] For an analysis of the use of various forms of de-monstration (including demos) in the construction of Europe, and more especially in the management of the major research and development programs of the European Union, see Claude Rosental, "Cooperation, Competition and Demonstration in the European Telecommunication Industry: A Sociological Analysis of the Development of Information Super Highways in Europe," Brunel University, Uxbridge, U.K., CRICT Paper no. 61, 1998.

WORKS CITED

Agre, Philip, and David Chapman. "Pengi: An Implementation of a Theory of Activity." In *Proceedings of the Sixth National Annual Conference on Artificial Intelligence*, 261–72. Los Altos, Calif., 1987.

Alker, Hayward R. "Standard Logic versus Dialectical Logic: Which Is Better for Scientific Political Discourse?" Paper presented at the Twelfth World Congress of the International Political Science Association, Rio de Janeiro, August 8–15, 1982.

Andler, Daniel. "Calcul et représentation: Les sources." In *Introduction aux sciences cognitives*, edited by Daniel Andler, 9–46. Paris: Gallimard, 1992.

Ascher, Marcia. *Ethnomathematics: A Multicultural View of Mathematical Ideas*. Pacific Grove, Calif.: Brooks/Cole, 1991.

Ashmore, Malcolm. "The Theatre of the Blind: Starring a Promethean Prankster, a Phoney Phenomenon, a Prism, a Pocket, and a Piece of Wood." *Social Studies of Science* 23 (1993): 67–106.

Aspray, William. *John von Neumann and the Origins of Modern Computing*. Cambridge: MIT Press, 1990.

Baldwin, J. F., and N.C.F. Guild. "The Resolution of Two Paradoxes by Approximate Reasoning Using a Fuzzy Logic." *Synthese* 44 (1980): 397–420.

Barnes, Barry, and Steven Shapin. *Natural Order: Historical Studies of Scientific Culture*. Beverly Hills: Sage, 1979.

Barwise, Jon, and John Etchemendy. *The Liar: An Essay on Truth and Circularity*. Oxford: Oxford University Press, 1987.

Barwise, Jon, and Solomon Feferman, eds. *Model-Theoretic Logics*. New York: Springer, 1985.

Batali, John. "The Power and the Story." *Postmodern Culture* 2, no. 2 (1992).

Belohlavek, Rahim, and George J. Klir. "On Elkan's Theorems: Clarifying Their Meaning via Simple Proofs." *International Journal of Intelligent Systems* 22, no. 2 (2007): 203–7.

Berenji, Hamid R. "The Unique Strength of Fuzzy Logic Control." *IEEE Expert*, August 1994, 9.

Berenji, Hamid R., James C. Bezdek, Pietro P. Bonissone, Didier Dubois, Rudolf Kruse, Henri Prade, Philippe Smets, and Ronald R. Yager. "A Reply to the Paradoxical Success of Fuzzy Logic." *AI Magazine*, Spring 1994, 6–8.

Bergmann, Merrie, James Moor, and Jack Nelson. *The Logic Book*. New York: Random House, 1980.

Berthelot, J.-M. *Les vertus de l'incertitude: Le travail de l'analyse dans les sciences sociales*. Paris: Presses Universitaires de France, 1996.

Biagioli, Mario. *The Science Studies Reader*. London: Routledge, 1999.

Black, Max. "Vagueness: An Exercise in Logical Analysis." *Philosophy of Science* 4, no. 4 (1937): 427–55.

Bloor, David. "Hamilton and Peacock on the Essence of Algebra." In *Social History of Nineteenth Century Mathematics*, edited by Herbert Mehrtens, Henk Bos, and Ivo Schneider, 202–32. Boston: Birkhauser, 1981.

———. *Knowledge and Social Imagery.* 2nd ed. Chicago: University of Chicago Press, 1991.

———."Polyhedra and the Abominations of Leviticus." *British Journal for the History of Science* 2, no. 39 (1978): 245–72.

———. "Polyhedra and the Abomination of Leviticus: Cognitive Styles in Mathematics." In *Essays in the Sociology of Perception*, edited by Mary Douglas, 191–218. London: Routledge and Kegan Paul, 1982.

———. *Wittgenstein: A Social Theory of Knowledge.* London: Macmillan, 1983.

———. "Wittgenstein and Mannheim on the Sociology of Mathematics." *Studies in the History and Philosophy of Science* 4 (1973): 173–91.

Blumenberg, Hans. *The Genesis of the Copernican World.* Cambridge: MIT Press, 1987.

Boll, Marcel, and Jacques Reinhart. *Histoire de la logique.* Paris: Presses Universitaires de France, 1946.

Boltanski, Luc, and Laurent Thévenot. *On Justification.* Translated by Catherine Porter. Princeton: Princeton University Press, 2006 (1991).

Boole, George. *The Laws of Thought.* New York: Dover, 1958.

Bouchon-Meunier, Bernadette. *La logique floue.* Paris: Presses Universitaires de France, 1993.

Boudon, Raymond. *The Art of Self-Persuasion: The Social Explanation of False Beliefs.* Translated by Malcolm Slater. Cambridge, Mass.: Polity Press, 1994.

Bourdieu, Pierre. *Homo Academicus.* Translated by Peter Collier. Stanford, Calif.: Stanford University Press, 1988.

———. *The State Nobility: Elite Schools in the Field of Power.* Translated by Lauretta C. Clough. Stanford, Calif.: Stanford University Press, 1996.

Boureau, Alain. "L'adage *Vox Populi, Vox Dei* et l'invention de la nation anglaise (VIIIe-XIIe siècle)." *Annales ESC*, no. 4–5 (1992): 1071–89.

———. "Propositions pour une histoire restreinte des mentalités." *Annales ESC* 6 (1989): 1491–504.

Bouveresse, Jacques. *La force de la règle.* Paris: Minuit, 1987.

———. *Le mythe de l'intériorité: Expérience, signification et langage privé chez Wittgenstein.* Paris: Minuit, 1976.

———. *Prodiges et vertiges de l'analogie: De l'abus des belles-lettres dans la pensée.* Paris: Raisons d'agir, 1999.

Bowers, John. "The Politics of Formalism." In *Contexts of Computer Mediated Communication*, edited by Martin Lea, 232–61. Hassocks: Harvester, 1992.

Boyer, Carl B. *The History of the Calculus and Its Conceptual Development.* New York: Dover, 1959.

Brian, Éric. "Le livre des sciences est-il écrit dans la langue des historiens?" In *Les formes de l'expérience: Une autre histoire sociale*, edited by Bernard Lepetit, 85–98. Paris: Albin Michel, 1995.

———. *La mesure de l'État: Administrateurs et géomètres au XVIIIe siècle.* Paris: Albin Michel, 1994.

Brouwer, L.E.J. "Historical Background, Principles and Methods of Intuition-ism." *South African Journal of Science* 49 (1952): 139–46.

Callon, Michel. "Some Elements of a Sociology of Translation: Domestication of the Scallops and the Fishermen of St. Brieuc Bay." In *Power, Action and Belief: A New Sociology of Knowledge?* edited by John Law, 196–223. London: Routledge, 1986.

Cambrosio, Alberto, Peter Keating, Thomas Schlich, and George Weisz. "Regulatory Objectivity and the Generation and Management of Evidence in Medicine." *Social Science and Medicine* 63 (2006): 189–99.

Canguilhem, Georges. *Le normal et le pathologique.* Paris: Presses Universitaires de France, 1972.

Carnap, Rudolph. *The Logical Syntax of Language.* Translated by Amethe Sweaton. London: K. Paul, Trench, Trubner, 1937.

Cassin, Barbara. *L'effet sophistique.* Paris: Gallimard, 1995.

Cazeneuve, Jean. *Lucien Lévy-Bruhl.* Paris: Presses Universitaires de France, 1963.

Chalvon-Demersay, Sabine. "Une société élective: Scénarios pour un mode de relations choisies." *Terrain* 27 (1996): 81–100.

Chartier, Roger, ed. *Les usages de l'imprimé: XVe–XIXe siècle.* Paris: Fayard, 1987.

Chemla, Karine. "Histoire des sciences et matérialité des textes: Proposition d'enquête." *Enquête: Anthropologie, histoire, sociologie* 1 (1995): 167–80.

Churchland, P. M. *A Neurocomputational Perspective: The Nature of Mind and the Structure of Science.* Cambridge: MIT Press, 1989.

Cifoletti, Giovanna Cleonice. "Mathematics and Rhetoric: Pelletier and Gosselin and the Making of the French Algebraic Tradition." Ph.D. diss., Princeton University, 1992.

Clark, Adele E., and Joan H. Fujimura, eds. *The Right Tools for the Job: At Work in Twentieth-Century Life Sciences.* Princeton: Princeton University Press, 1992.

Clark, R. N. *Introduction to Automatic Control Systems.* New York: John Wiley, 1962.

Claverie, Élisabeth. *Les guerres de la vierge: Une anthropologie des apparitions.* Paris: Gallimard, 2003.

Cole, Michael, and Sylvia Scribner. *Culture and Thought: A Psychological Introduction.* New York: Wiley and Sons, 1974.

Coleman, Edwin. "The Role of Notation in Mathematics." Ph.D. diss., University of Adelaide, Australia, 1988.

Collins, Harry M. *Artificial Experts: Social Knowledge and Intelligent Machines.* Cambridge: MIT Press, 1990.

———. *Changing Order: Replication and Induction in Scientific Practice.* London: Sage, 1985.

Collins, Randall, and Sal Restivo. "Robber Barons and Politicians in Mathematics: A Conflict Model of Science." *Canadian Journal of Sociology* 8 (1983): 199–227.

Coulon, Alain. *Ethnométhodologie et éducation.* Paris: Presses Universitaires de France, 1993.

Cowan, J. D., and H. D. Sharp. "Neural Nets and Artificial Intelligence." In *The Artificial Intelligence Debate: False Starts, Real Foundations*, edited by S. R. Graubard, 85–121. Cambridge: MIT Press, 1989.

Croarken, Mary. *Early Scientific Computing in Britain*. Oxford: Clarendon Press, 1990.

Crump, Thomas. *The Anthropology of Numbers*. Cambridge: Cambridge University Press, 1990.

Cussins, Adrian. "Content, Embodiment and Objectivity: The Theory of Cognitive Trails." *Mind* 101, no. 404 (1992): 651–88.

Dahan-Dalmedico, Amy. "Réponse à Hélène Gispert." In *Le relativisme est-il résistible? Regards sur la sociologie des sciences*, edited by Raymond Boudon and Maurice Clavelin, 221–25. Paris: Presses Universitaires de France, 1994.

Dahan-Dalmedico, Amy, and Jeanne Pfeiffer. *Une histoire des mathématiques: Routes et dédales*. Paris: Seuil, 1986.

Daston, Lorraine. "Marvelous Facts and Miraculous Evidence in Early Modern Europe." *Critical Inquiry* 19 (1991): 93–124.

———. "Objectivity and the Escape from Perspective." *Social Studies of Science* 2 (1992): 597–618.

Davis, Philip J., and Reuben Hersh. *Descartes' Dream: The World according to Mathematics*. San Diego: Harcourt Brace Jovanovich, 1986.

———. *The Mathematical Experience*. Boston: Birkhauser, 1981.

———. "Rhetoric and Mathematics." In *The Rhetoric of the Human Sciences: Language and Argument in Scholarship and Public Affairs*, edited by John S. Nelson, Alan Megill, and Donald N. McCloskey, 53–68. Madison: University of Wisconsin Press, 1987.

Dedekind, Richard. *Essays on the Theory of Numbers*. New York: Dover, 1963.

De Millo, Richard A., Richard J. Lipton, and Alan J. Perlis. "Social Processes and Proofs of Theorems and Programs." *Communications of the Association for Computing Machinery* 22, no. 5 (1979): 271–80.

Dennett, Daniel C. *Consciousness Explained*. Boston: Little, Brown, 1991.

Doppelt, Gerald. "Kuhn's Epistemological Relativism: An Interpretation and Defense." *Inquiry* 21 (1978): 33–86.

Douglas, Mary. *Natural Symbols*. London: Barrie and Jenkins, 1970.

Dubois, Didier, and Henri Prade. "Can We Enforce Full Compositionality in Uncertainty Calculi?" *Proceedings of the 1994 National Conference on Artificial Intelligence*, 149–54. Cambridge: American Association for Artificial Intelligence, MIT Press, 1994.

———. "An Introduction to Possibilistic and Fuzzy Logics." In *Nonstandard Logics for Automated Reasoning*, edited by Philippe Smets et al. London: Academic Press, 1988.

Dubois, Didier, Henri Prade, and Philippe Smets. "Partial Truth Is Not Uncertainty: Fuzzy Logic versus Possibilistic Logic." *IEEE Expert*, August 1994, 15–19.

Dulong, Renaud. *Le témoin oculaire: Les conditions sociales de l'attestation personnelle*. Paris: Éditions de l'École des Hautes Études en Sciences Sociales, 1998.

Dummett, Michael A. E., ed. *Truth and Other Enigmas*. Cambridge: Harvard University Press, 1978.

Dupuy, Jean-Pierre. *Mechanization of the Mind: On the Origins of Cognitive Science*. Translated by M. B. DeBevoise. Princeton: Princeton University Press, 2000.

Durkheim, Émile. *The Elementary Forms of Religious Life*. Translated by Joseph Ward Swain. 2nd ed. London: George Allen and Unwin, 1976.

Eisenstein, Elizabeth. *The Printing Press as an Agent of Change: Communication and Cultural Transformations in Early Modern Europe*. Cambridge: Cambridge University Press, 1972.

Elkan, Charles. "Conspiracy Numbers and Caching for Searching and/or Trees and Theorem Proving." In *Proceedings of the Eleventh International Joint Conference on Artificial Intelligence*, August 1989, 341–48. San Francisco: Morgan Kaufmann, 1989.

———. "Formalizing Causation in First-Order Logic: Lessons from an Example." Working notes of the AAAI Spring Symposium on Logical Formalizations of Commonsense Reasoning. 41–47. Stanford University, March, 1991.

———. "Incremental, Approximate Planning." In *Proceedings of the Eighth National Conference on Artificial Intelligence*, 145–50. Boston: AAAI Press, MIT Press, 1990.

———. "Paradoxes of Fuzzy Logic, Revisited: Discussion." *International Journal of Approximate Reasoning* 26, no. 2 (2001): 153–55.

———. "The Paradoxical Success of Fuzzy Logic." In *Proceedings of the Eleventh National Conference on Artificial Intelligence*, 698–703. Menlo Park, Calif.: AAAI Press; Cambridge: MIT Press, 1993.

———. "The Paradoxical Success of Fuzzy Logic." cs.ucsd.edu/pub/paradoxical success.ps, November 1993.

———. "The Paradoxical Success of Fuzzy Logic." *IEEE Expert*, August 1994, 3–8.

———. "The Qualification Problem, the Frame Problem, and Nonmonotonic Logic." Working paper, University of California, San Diego, 1992.

———. "Reasoning about Action in First Order Logic." In *Proceedings of the Ninth Biennial Conference of the Canadian Society for Computational Studies of Intelligence*, 221–27.

———. "Research on the Qualification and Frame Problems." Miller Research Professorship Application, University of California, San Diego, January 1992.

———. "The Theory in My Paper Is Correct . . ." *AI Magazine*, Spring 1994, 8.

Elkan, Charles, and David A. McAllester. "Automated Inductive Reasoning about Logic Programs." In *Proceedings of the Fifth International Conference Symposium on Logic Programming*, ed. Kenneth A. Bowen and Robert Kowalski, 876–92. Cambridge: MIT Press, 1988.

Entemann, Carl W. "Fuzzy Logic: Misconceptions and Clarifications." *Artificial Intelligence Review* 17, no. 1 (2002): 65–84.

Erdmann, Benno. *Logik*. Berlin: W. de Gruyter, 1923.

Evans-Pritchard, E. E. *Witchcraft, Oracles and Magic among the Azande*. Oxford: Clarendon Press, 1937.

Fabiani, Jean-Louis. *Les philosophes de la République*. Paris: Minuit, 1988.

Fisher, Charles S. "The Death of a Mathematical Theory: A Study in the Sociology of Knowledge." *Archives for History of Exact Sciences* 3 (1966): 137–59.

———. "Some Characteristics of Mathematicians and Their Work." *American Journal of Sociology* 78 (1973): 1094–118.

Fleck, Ludwig. *Genesis and Development of a Scientific Fact*. Translated by Fred Bradley and Thaddeus J. Trenn. Chicago: University of Chicago Press, 1979.

Fox, John. "Towards a Reconciliation of Fuzzy Logic and Standard Logic." *International Journal of Man-Machine Studies* 15 (1981): 213–20.

Frege, Gottlob. *The Foundations of Arithmetic: A Logico-mathematical Enquiry into the Concept of Number*. Translated by J. L. Austin. 2nd ed. New York: Philosophical Library, 1953.

———. *Translations from the Philosophical Writings of Gottlob Frege*. Edited by Peter Geach and Max Black. Oxford: Blackwell, 1959.

Friedman, Robert Marc. *Appropriating the Weather*. Ithaca, N.Y.: Cornell University Press, 1989.

Gaines, Brian R. "Precise Past, Fuzzy Future." *International Journal of Man-Machine Studies* 19 (1983): 117–34.

Gaines, Brian R., and L. Kohout. "The Fuzzy Decade: A Bibliography of Fuzzy Systems and Closely Related Topics." *International Journal of Man-Machine Studies* 9 (1977): 1–68.

Galison, Peter, and David J. Stump. *The Disunity of Science: Boundaries, Contexts, and Power*. Stanford, Calif.: Stanford University Press, 1996.

Garfinkel, Harold. *Studies in Ethnomethodology*. Englewood Cliffs, N.J.: Prentice Hall, 1968.

Gigerentzer, Gerd. "From Tools to Theories: A Heuristic of Discovery in Cognitive Psychology." *Psychological Review* 98, no. 2 (1991): 254–67.

Gingras, Yves. "From the Heights of Metaphysics: A Reply to Pickering." *Social Studies of Science* 29, no. 2 (1999): 312–26.

———. "What Did Mathematics Do to Physics?" *History of Science* 39, no. 126 (2001): 383–416.

Girard, J. Y. "Linear Logic." *Theoretical Computer Science* 50 (1987): 1–102.

Girotto, Vittorio. "Judgements of Deontic Relevance in Reasoning: A Reply to Jackson and Griggs." *Quarterly Journal of Experimental Psychology* 45A, no. 4 (1992): 547–74.

Gispert, Hélène. "Un exemple d'approche sociologique en histoire des mathématiques: L'analyse au XIXe siècle." In *Le relativisme est-il résistible? Regards sur la sociologie des sciences*, edited by Raymond Bourdon and Maurice Clavelin, 211–20. Paris: Presses Universitaires de France, 1994.

———. *La France mathématique: La Société mathématique de France (1872–1914)*. Paris: Société française des sciences et des techniques, Société mathématique de France, 1991.

Goffman, Erving. *Forms of Talk*. Philadelphia: University of Pennsylvania Press, 1981.

———. *The Presentation of Self in Everyday Life*. Garden City, N.Y.: Doubleday, 1959.

Goguen, J. A. "The Logic of Inexact Concepts." *Synthese* 19 (1969): 325–73.

Goldstein, Catherine. *Un théorème de Fermat et ses lecteurs*. Saint-Denis: Presses Universitaires de Vincennes, 1995.

———. "Working with Numbers in the Seventeenth and Nineteenth Centuries." In *A History of Scientific Thought: Elements of a History of Science*, edited by Michel Serres, 344–71. Oxford: Blackwell, 1995.

Goldstine, Herman Heine. *The Computer from Pascal to von Neumann*. Princeton: Princeton University Press, 1972.

Gooding, David, Trevor Pinch, and Simon Schaffer, eds. *The Uses of Experiment: Studies in the Natural Sciences*. Cambridge: Cambridge University Press, 1989.

Goody, Jack. *The Domestication of the Savage Mind*. Cambridge: Cambridge University Press, 1977.

Granet, Marcel. *Études sociologiques sur la Chine*. Paris: Presses Universitaires de France, 1953.

Haack, Susan. "Do We Need 'Fuzzy Logic'?" *International Journal of Man-Machine Studies* 11 (1979): 437–45.

Hacking, Ian. *The Social Construction of What?* Cambridge: Harvard University Press, 1999.

Hamill, James F. *Ethno-logic: The Anthropology of Human Reasoning*. Urbana: University of Illinois Press, 1990.

Hardy, Godfrey Harold. "Mathematical Proof." *Mind* 38 (1929): 1–25.

Harel, Idit, and Seymour Papert. "Software Design as a Learning Environment." *Interactive Learning Environments* 1 (1990): 1–32.

Hay, Cynthia, ed. *Mathematics from Manuscript to Print: 1500–1600*. Oxford: Clarendon Press, 1988.

Hennion, Antoine. *Comment la musique vient aux enfants: Une anthropologie de l'enseignement musical*. Paris: Anthropos, 1988.

———. *La passion musicale*. Paris: Métailié, 1993.

Hilbert, David. "Axiomatisches Denken." *Mathematische Annalen* 78 (1918): 405–15.

Hilbert, David, and Wilhelm Ackermann. *Grundzüge der theoretischen Logik*. Berlin: J. Springer, 1928.

Hodges, Andrew. *Alan Turing: The Enigma*. New York: Simon and Schuster, 1983.

Hodgkin, Luke. "Mathematics as Ideology and Politics." In *Radical Science Essays*, edited by Les Levidow, 173–97. London: Free Association Books, 1986.

Horton, Robin. "Tradition and Modernity Revisited." In *Rationality and Relativism*, edited by Martin Hollis and Steven Lukes, 201–60. Cambridge: MIT Press, 1982.

Horton, Robin, and Ruth Finnegan. *Modes of Thought: Essays on Thinking in Western and Non-Western Societies*. London: Faber, 1973.

Husserl, Edmund. *Logical Investigations*. Translated by J. N. Findlay. London: Routledge, 2001.

Hutchins, Edwin. *Cognition in the Wild*. Cambridge: MIT Press, 1995.

IEEE Expert. Special issue. August 1994.

Iliffe, Robert. "Theory, Experiment, and Society and French and Anglo-Saxon History of Science." *Revue européenne d'histoire* 2, no. 1 (1995): 73–74.

Ionesco, Eugene. "The Lesson." In *Four Plays*, edited by Donald M. Allen. New York: Grove Press, 1958.

Jacq, François. "Pratiques scientifiques, formes d'organisation et représentations politiques de la science dans la France d'après-guerre: La 'politique de la science' comme énoncé collectif." Ph.D. diss., École Nationale Supérieure des Mines de Paris, 1996.

Jasanoff, Sheila, et al. *Handbook of Science and Technology Studies*. London: Sage, 1995.

Jullien, François. *The Propensity of Things: Toward a History of Efficacity in China*. Translated by Janet Lloyd. New York: Zone, 1995.

Kitcher, Philip. *The Nature of Mathematical Knowledge*. New York: Oxford University Press, 1984.

Kline, Morris. *Mathematics: The Loss of Certainty*. New York: Oxford University Press, 1980.

Knowles, Greg. *An Introduction to Applied Optimal Control*. New York: Academic Press, 1981.

Kosko, Bart. "Fuzziness vs. Probability." *International Journal of General Systems* 17 (1990): 211–40.

———. *Fuzzy Thinking: The New Science of Fuzzy Logic*. New York: Hyperion, 1993.

———. *Neural Networks and Fuzzy Systems: A Dynamical Systems Approach to Machine Intelligence*. Englewood Cliffs, N.J.: Prentice Hall, 1992.

Kosko, Bart, and S. Isaka. "Fuzzy Logic." *Scientific American*, July 1993, 76–81.

Kuhn, Thomas. *The Structure of Scientific Revolutions*. Chicago: University of Chicago Press, 1962.

Kuntzmann, Jean. *Fundamental Boolean Algebra*. Translated by Scripta Technica Ltd. London: Blackie, 1967.

Lahire, Bernard. *L'homme pluriel*. Paris: Nathan, 1998.

Lakatos, Imre. *Proofs and Refutations: The Logic of Mathematical Discovery*. Edited by John Worrall and Elie Zahar. Cambridge: Cambridge University Press, 1976.

Lamping, J. O. "A Unified System of Parametrization for Programming Language." Ph.D. diss., Stanford University, 1988.

Largeault, Jean. *L'intuitionnisme*. Paris: Presses Universitaires de France, 1992.

———. *La logique*. Paris: Presses Universitaires de France, 1993.

Latour, Bruno. "Cogito Ergo Sumus! Or Psychology Swept Inside Out by the Fresh Air of the Upper Deck . . ." *Mind, Culture, and Activity* 3, no. 1 (1996): 54–68.

———. "Give Me a Laboratory and I Will Raise the World." In *Science Observed: Perspectives on the Social Studies of Science*, edited by Karin D. Knorr, and Michael Mulkay, 141–70. London: Sage, 1983.

———. "Les idéologies de la compétence en milieu industriel à Abidjan." *Cahiers Orstom—Sciences humaines*, no. 9 (1973): 1–174.

———. "One More Turn after the Social Turn: Easing Science Studies into the Non-modern World." In *The Social Dimensions of Science*, edited by Ernan McMullin, 272–94. Notre Dame, Ind.: Notre Dame University Press, 1992.

————. *The Pasteurization of France.* Translated by Alan Sheridan and John Law. Cambridge: Harvard University Press, 1988.

————. *Petite réflexion sur le culte moderne des dieux Faitiches.* Le Plessis-Robinson: Synthélabo, 1996.

————. *Petites leçons de sociologie des sciences.* Paris: La Découverte, 1993.

————. *Science in Action: How to Follow Scientists and Engineers through Society.* Cambridge: Harvard University Press, 1987.

————. "Sur la pratique des théoriciens." In *Savoirs théoriques et savoirs d'action,* edited by J.-M. Barbier, 131–46. Paris: Presses Universitaires de France, 1996.

————. "Visualisation and Cognition: Seeing with Eyes and Hands." *Knowledge and Society* 6 (1986): 1–40.

Latour, Bruno, and Steve Woolgar. *Laboratory Life: The Social Construction of Scientific Facts.* Preface by Jonas Salk. Princeton: Princeton University Press, 1986 (1979).

————. *La vie de laboratoire.* Translated by Michel Biezunski. Paris: La Découverte, 1988.

Lave, Jean. *Cognition in Practice: Mind, Mathematics and Culture in Everyday Life.* Cambridge: Cambridge University Press, 1988.

Lave, Jean, James Greeno, Alan Schoenfeld, Steven Smith, and Michael Butler. "Learning Mathematical Problem Solving." Palo Alto, Calif.: Institute for Research on Learning, Report no. IRL88–0006. 1988.

Lebesgue, Henri Léon. *Leçons sur l'intégration et la recherche des functions primitives.* Paris: Gauthier-Villars, 1928.

Lévy, Pierre. *De la programmation considérée comme un des beaux-arts.* Paris: La Découverte, 1992.

Lévy-Bruhl, Lucien. *Primitive Mentality.* Translated by Lilian A. Clare. Boston: Beacon Press, 1966.

Licoppe, Christian. *La formation de la pratique scientifique: Le discours de l'expérience en France et en Angleterre (1630–1820).* Paris: La Découverte, 1996.

Livingston, Eric. "Cultures of Proving." *Social Studies of Science* 29 (1999): 867–88.

————. *The Ethnomethodological Foundations of Mathematics.* London: Routledge, 1985.

Lloyd, G.E.R. *Demystifying Mentalities.* Cambridge: Cambridge University Press, 1990.

Lukasiewicz, Jan. *Selected Works.* Edited by Ludvik Borkowski. Amsterdam: North Holland, 1970.

Lynch, Michael. *Art and Artifact in Laboratory Science: A Study of Shop Work and Shop Talk in a Research Laboratory.* London: Routledge and Kegan Paul, 1985.

————. "Discipline and the Material Form of Images: An Analysis of Scientific Visibility." *Social Studies of Science* 15 (1985): 37–66.

————. "The Externalized Retina: Selection and Mathematization in the Visual Documentation of Objects in the Life Sciences." *Human Studies* 11 (1998): 201–34.

Lynch, Michael, and Steve Woolgar, eds. *Representation in Scientific Practice.* Cambridge: MIT Press, 1990.

MacKenzie, Donald. "The Fangs of the VIPER." *Nature* 352 (Aug. 8, 1991): 467–68.

———. *Mechanizing Proof: Computing, Risk, and Trust.* Cambridge: MIT Press.

———. "Negotiating Arithmetic, Constructing Proof: The Sociology of Mathematics and Information Technology." *Social Studies of Science* 23 (1993): 37–65.

———. *Statistics in Britain, 1865–1930: The Social Construction of Scientific Knowledge.* Edinburgh: Edinburgh University Press, 1981.

Mannheim, Karl. *Essays on the Sociology of Knowledge.* London: Routledge, 1952.

McNeill, Dan, and Paul Freiberger. "The Authors Respond." *IEEE Spectrum,* September 1993, 6, 8.

———. *Fuzzy Logic: The Discovery of a Revolutionary Computer Technology, and How It Is Changing Our World.* New York: Simon and Schuster, 1992.

———. "Out of Focus on Fuzzy Logic." *IEEE Spectrum,* August 1993, 6.

Mehrtens, Herbert, Henk Bos, and Ivo Schneider, eds. *Social History of Nineteenth Century Mathematics.* Boston: Birkhauser, 1981.

Menger, Pierre-Michel. *Les laboratoires de la création musicale: Acteurs, organisations et politique de la recherche musicale.* Paris: La Documentation française, 1989.

Mialet, Hélène. "Le sujet de l'invention." Ph.D. diss., Université de Paris I–Sorbonne, 1994.

Mill, John Stuart. *A System of Logic, Ratiocinative and Inductive: Being a Connected View of the Principles of Evidence and the Methods of Scientific Investigation.* London: Longmans, 1961.

Mitchell, Melanie. *An Introduction to Genetic Algorithms.* Cambridge: MIT Press, 1996.

Mukerji, Chandra. *A Fragile Power: Scientists and the State.* Princeton: Princeton University Press, 1989.

———. *Territorial Ambitions and the Gardens of Versailles.* Cambridge: Cambridge University Press, 1967.

Naville, Pierre. *Sociologie et logique: Esquisse d'une théorie des relations.* Paris: Presses Universitaires de France, 1982.

Nye, Andrea. *Words of Power: A Feminist Reading in the History of Logic.* London: Routledge, 1990.

Ochs, Elinor, Sally Jacoby, and Patrick Gonzales. "Interpretive Journeys: How Physicists Talk and Travel through Graphic Space." *Configurations* 2, no. 1 (1994): 151–71.

Pan, Juiyao, Guilherme N. DeSouza, and Avinash C. Kak. "FuzzyShell: A Large-Scale Expert System Shell Using Fuzzy Logic for Uncertainty Reasoning." *IEEE Transactions on Fuzzy Systems* 6, no. 4 (1998): 563–81.

Passeron, Jean-Claude. *Le raisonnement sociologique: L'espace non poppérien du raisonnement naturel.* Paris: Nathan, 1991.

Perrot, Jean-Claude. *Une histoire intellectuelle de l'économie politique (XVII-XVIIIe siècles).* Paris: École des Hautes Études en Sciences Sociales, 1992.

Pestre, Dominique. *Physique et physiciens en France, 1918–1946*. Paris: Archives Contemporaines, 1984.

———. "Pour une histoire sociale et culturelle des sciences: Nouvelles définitions, nouveaux objets, nouvelles pratiques." *Annales Histoire, Sciences Sociales* 3 (1995): 504–6.

Pickering, Andrew. "In the Land of the Blind . . . Thoughts on Gingras." *Social Studies of Science* 29, no. 2 (1999): 307–11.

———. *Mangle of Practice: Time, Agency and Science*. Chicago: University of Chicago Press, 1995.

Pickering, Andrew, and Adam Stephanides. "Constructing Quaternions: On the Analysis of Conceptual Practice." In *Science as Practice and Culture*, edited by Andrew Pickering, 139–67. Chicago: University of Chicago Press, 1992.

Polanyi, Michael. *Personal Knowledge: Towards a Post-Critical Philosophy*. London: Routledge, 1958.

———. *The Tacit Dimension*. New York: Anchor, 1958.

Politzer, Guy, ed. "Pragmatique et psychologie du raisonnement." Vol. 11, *Intellectica* (special issue), 1991.

Polya, George. *Mathematical Discovery: On Understanding, Learning, and Teaching Problem Solving*. 2 vols. New York: Wiley, 1972.

Popper, Karl R. *The Logic of Scientific Discovery*. London: Routledge, 2002.

Porridge, Pease. "What's All This Fuzzy Logic Stuff, Anyhow?" *Electronic Design*, May 13, 1993, 77–79.

Proust, Marcel. *La Captive*. In *Remembrance of Things Past*. Trans. C. K. Scott Moncrieff, 383–669. Vol. 2. New York: Random House, 1927.

Puppe, Frank. *Systematic Introduction to Expert Systems: Knowledge Representations and Problem Solving Methods*. New York: Springer-Verlag, 1993.

Quéré, Louis. "Agir dans l'espace public: L'intentionnalité des actions comme phénomène social." In *Les formes de l'action*, edited by Patrick Pharo and Louis Quéré, 85–112. Paris: EHESS, 1990.

Quine, Willard Van Orman. "Truth by Convention." In *The Ways of Paradox and Other Essays*, 70–99. Cambridge: Harvard University Press, 1966. Reprint, 1976.

———. *The Ways of Paradox and Other Essays*. Cambridge: Harvard University Press, 1966. Reprint, 1976.

Ravetz, Jerome R. *Scientific Knowledge and Its Social Problems*. Oxford: Clarendon Press, 1971.

Rawlins, G. J. E., ed. *Foundations of Genetic Algorithms*. San Mateo, Calif.: Morgan Kaufmann, 1991.

Restivo, Sal. *Mathematics in Society and History: Sociological Inquiries*. Dordrecht: Kluwer, 1992.

———. *The Social Relations of Physics, Mysticism, and Mathematics*. Dordrecht: D. Riedel, 1983.

Restivo, Sal, Jean Paul von Bendegem, and Roland Fischer. *Math Worlds: Philosophical and Social Studies of Mathematics and Mathematics Education*. Albany: State University of New York Press, 1991.

Ritter, James. "Les pratiques rationnelles en Égypte et en Mésopotamie aux troisième et deuxième millénaires." Ph.D. diss., Université de Paris XIII–Villetaneuse, 1993.

Rosental, Claude. "The Art of Monstration and Targeted Doubt: The Anthropology of Mathematical Demonstration." Working paper, Centre Sociologie de l'Innovation, École des Mines de Paris, 1993.

———. *Les capitalistes de la science: Enquête sur les démonstrateurs de la Silicon Valley et de la NASA.* Paris: CNRS Éditions, 2007.

———. *Cooperation, Competition and Demonstration in the European Telecommunication Industry: A Sociological Analysis of the Development of Information Super Highways in Europe.* Uxbridge: CRICT Paper no. 61, 1998.

———. "De la demo-cratie en Amérique: formes actuelles de la démonstration en intelligence artificielle." *Actes de la recherche en sciences sociales* 141–42 (2002): 110–20.

———. "De la logique au pathologique." DEA diss., Université de Paris I–Sorbonne, 1992.

———. "An Ethnography of the Teaching of Logic." Paper presented at the Workshop on Historical Epistemology, Institute for the History of Science, University of Toronto, 1993.

———. "Fuzzyfying the World: Social Practices of Showing the Properties of Fuzzy Logic." In *Growing Explanations: Historical Perspectives on Recent Science*, edited by M. Norton Wise, 279–319. Durham, N.C.: Duke University Press, 2004.

———. "Histoire de la logique floue: Une approche sociologique des pratiques de démonstration." *Revue de Synthèse* 4, no. 4 (1998): 573–602.

———. "Making Science and Technology Results Public: A Sociology of Demos." In *Making Things Public: Atmospheres of Democracy*, edited by Bruno Latour and Peter Weibel, 346–49. Cambridge: MIT Press, 2005.

———. "Quelle logique pour quelle rationalité? Représentations et usages de la logique en sciences sociales." In *L'argumentation: Preuve et persuasion*, edited by Michel de Fornel and Jean-Claude Passeron, 69–92. Enquête, vol. 2. Paris: Éditions de l'École des Hautes Études en Sciences Sociales, 2002.

———. "Le rôle d'Internet dans l'évolution des pratiques, des formes d'organisation, et des réseaux de la recherche." *Annales des Mines* (Feb. 1998): 103–8.

———. "Les travailleurs de la preuve sur Internet: Transformations et permanences du fonctionnement de la recherche." *Actes de la recherche en sciences sociales* 134 (2000): 37–44.

Rota, Gian-Carlo. "The Pernicious Influence of Mathematics upon Philosophy." *Synthese* 88 (1991): 165–78.

Rotman, Brian. *Ad Infinitum—the Ghost in Turing's Machine. Taking God Out of Mathematics and Putting the Body Back In: An Essay in Corporeal Semiotics.* Stanford, Calif.: Stanford University Press, 1993.

———. *Signifying Nothing: The Semiotics of Zero.* London: Macmillan, 1987.

———. "Thinking Dia-grams: Mathematics, Writing, and Virtual Reality." *South Atlantic Quarterly* 94, no. 2 (1995): 389–415.

Ruspini, Enrique H. "Fuzzy Logic Revisited." *IEEE Spectrum*, September 1993, 6.

———. "On the Purportedly Paradoxical Nature of Fuzzy Logic." *IEEE Expert*, August 1994, 32–33.

———. "An Opportunity Missed." *IEEE Spectrum*, June 1993, 11, 12, 15.

Russell, Bertrand, and Alfred North Whitehead. *Principia Mathematica.* Cambridge: Cambridge University Press, 1910.

Saffiotti, Alessandro, Nicholas Helft, Kurt Konolige, John Lowrance, Karen Myers, Daniela Musto, Enrique Ruspini, and Leonard Wesley. "A Fuzzy Controller for Flakey, the Robot." In *Proceedings of AAAI 1993,* 864. Cambridge: MIT Press, 1993.

Sainsbury, R. M. *Paradoxes.* Cambridge: Cambridge University Press, 1988.

Schaffer, Simon. "Self-Evidence." *Critical Inquiry* 18 (Winter 1992): 327–62.

———. "Universities, Instrument Shops and Demonstration Devices in 1776." Paper presented at Séminaire CRHST, "Spaces of Experiment." Paris, May 10, 1994.

Scheler, Max. *Die Wissenformen und die Gesellschaft.* Leipzig: Der neue Geist, 1926.

Schopenhauer, Arthur. *The World as Will and Idea.* Translated by R. B. Haldane and J. Kemp. 3 vols. London: Trübner, 1883–86.

Schütz, Alfred. "The Problem of Rationality in the Social World." *Economica* 10, no. 38 (1943): 130–49.

Sell, P. S. *Expert Systems: A Practical Introduction.* New York: John Wiley, 1985.

Serres, Michel. "Gnomon: The Beginnings of Geometry in Greece." In *A History of Scientific Thought: Elements of a History of Science,* edited by Michel Serres, 73–123. Oxford: Blackwell, 1995.

———. *Statues.* Paris: François Bourin, 1987.

Shapin, Steven. "Discipline and Bounding: The History and Sociology of Science as Seen through the Externalism-Internalism Debate." *History of Science* 30 (1992): 334–69.

———. "Robert Boyle and Mathematics: Reality, Representation and Experimental Practice." *Science in Context* 2, no. 1 (1988): 23–58.

———. *A Social History of Truth: Civility and Science in Seventeenth-Century England.* Chicago: University of Chicago Press, 1994.

Shapin, Steven, and Simon Schaffer. *Leviathan and the Air-Pump: Hobbes, Boyle and the Experimental Life.* Princeton: Princeton University Press, 1985.

Shu, Xiang Zhao, and Ping Lu Jian. "Measure-Based Fuzzy Operators Explain Elkan's Theorem." *IEEE Expert,* January 1996, 84–85.

Sigwart, Christoph. *Logik.* Tübingen: H. Laupp, 1889.

Snow, Paul. "What Concrete Things Does Fuzzy Propositional Logic Describe?" Eighth Pacific Rim International Conference on Artificial Intelligence, Auckland, New Zealand, August 9–13, 2004. *Proceedings Lecture Notes in Artificial Intelligence,* vol. 3157 (Berlin: Springer, 2004): 445–53.

Sossai, Claudio, Gaetano Chemello, and Alessandro Saffiotti, "A Note on the Role of Logic in Fuzzy Logic Controllers." *Proceedings of the Eighth International Conference on Information Processing and Management of Uncertainty in Knowledge-Based Systems,* 717–23. Madrid, 2000.

Sperber, Dan. "Les sciences cognitives, les sciences sociales et le matérialisme." In *Introduction aux sciences cognitives,* edited by Daniel Andler, 397–420. Paris: Gallimard, 1992.

Star, S. L., and J. R. Griesemer. "Institutional Ecology, 'Translations,' and Boundary Objects: Amateurs and Professionals in Berkeley's Museum of Vertebrate Zoology, 1907–1939." *Social Studies of Science* 19 (1989): 387–420.

Stark, Werner. *The Sociology of Knowledge: An Essay in Aid of a Deeper Understanding of the History of Ideas.* London: Routledge, 1958.

Suchman, Lucy A., and Randall H. Trigg. "Artificial Intelligence as Craftwork." In *Understanding Practice: Perspectives on Activity and Context*, edited by Seth Chaiklin and Jean Lave, 144–78. Cambridge: Cambridge University Press, 1993.

Synthèse technologies de l'information. Vol. 1. Paris: Aditech, 1989–90.

Trillas, Enric, and Claudi Alsina. "Comments to 'Paradoxes of Fuzzy Logic, Revisited: Discussion.' " *International Journal of Approximate Reasoning* 26, no. 2 (2001): 157–59.

———. "Elkan's Theoretical Argument, Reconsidered." *International Journal of Approximate Reasoning* 26, no. 2 (2001): 145–52.

Turner, Roy Steven. "The Growth of Professional Research in Prussia, 1818–1848: Causes and Contexts." *Historical Studies in the Physical Sciences* 3 (1971): 137–82.

Ulam, Stanislaw M. *Adventures of a Mathematician.* New York: Scribner, 1976.

Van Bendegem, Jean Paul. "Non-formal Properties of Real Mathematical Proofs." In *Proceedings of the 1988 Biennial Meeting of the Philosophy of Science Association*, edited by Arthur Fine, and Jarrett Leplin, 249–54. East Lansing, Mich.: Philosophy of Science Association, 1988.

Verran, Helen. *Science and an African Logic.* Chicago: University of Chicago Press, 2001.

Vinck, Dominique. *Sociologie des sciences.* Paris: Armand Colin, 1995.

Walton, D. N. *Arguer's Position: A Pragmatic Study of Ad Hominem Attack, Criticism, Refutation, and Fallacy.* Westport, Conn.: Greenwood Press, 1985.

Wang, Guo Jun, and Hao Wang. "Non-fuzzy Versions of Fuzzy Reasoning in Classical Logics." *Information Sciences* 138 (Oct. 2001): 211–36.

Warwick, Andrew. "Cambridge Mathematics and Cavendish Physics: Cunningham, Campbell and Einstein's Relativity Theory 1905–1911." Part 1, "The Uses of Theory." *Studies in the History and Philosophy of Science* 23, no. 4 (1992): 625–56.

———. "Cambridge Mathematics and Cavendish Physics: Cunningham, Campbell and Einstein's Relativity Theory 1905–1911." Part 2, "Comparing Traditions in Cambridge Physics." *Studies in the History and Philosophy of Science* 24, no. 1 (1993): 1–25.

———. "The Laboratory of Theory, or What's Exact about the Exact Sciences?" In *The Values of Precision*, edited by M. Norton Wise, 311–51. Princeton: Princeton University Press, 1995.

———. *Masters of Theory: Cambridge and the Rise of Mathematical Physics.* Chicago: University of Chicago Press, 2003.

Watkins, Fred A. "False Controversy: Fuzzy and Non-fuzzy Faux Pas." *IEEE Expert*, special issue, "Intelligent Systems and Their Applications," April 1995, 4–5.

Wilder, Raymond L. *Mathematics as a Cultural System.* Oxford: Pergamon, 1981.

Wilensky, Uri. "Abstract Meditations on the Concrete and Concrete Implications for Mathematics Education." Cambridge, Massachusetts Institute of Technology, Epistemology and Learning Group, Memo no. 12.

Winch, Peter. "Understanding a Primitive Society." *American Philosophical Quarterly* 1 (1964): 307–24.

Wise, M. Norton, ed. *The Values of Precision*. Princeton: Princeton University Press, 1995.

Wittgenstein, Ludwig. *Philosophical Investigations*. Translated by G.E.M. Anscombe. Oxford: Blackwell, 1953.

———. *Remarks on the Foundations of the Mathematics*. Translated by G.E.M. Anscombe. Oxford: Blackwell, 1956.

———. *Wittgenstein's Lectures on the Foundations of the Mathematics*. Edited by C. Diamond. Hassocks: Harvester, 1976.

Zadeh, L. A. "Fuzzy Logic." CSLI Report no. 88–116. Stanford, Calif., Stanford University, 1988.

———. "Fuzzy Sets." *Information and Control* 8 (1965): 338–53.

———. "Liar's Paradox and Truth-Qualification Principle." Memorandum no. UCB/ERL M79/34. University of California Berkeley, 1979.

———. "Why the Success of Fuzzy Logic Is Not Paradoxical." *IEEE Expert*, August 1994, 43–46.

Zimmermann, H.-J. *Fuzzy Set Theory—and Its Applications*. Dordrecht: Kluwer, 1985.

INDEX

Abstraction, 6, 35–36, 38, 49–51, 56

Access: to electronic forums, 56–58, 153; to Elkan's theorem, 10, 12–13, 109, 111–14; to field of study, 8, 14, 17, 109; to logic, 21, 71; to resources, 189, 204, 251

Accompaniment, 70, 156, 207, 209; actions of, 11, 14, 90–91, 113, 119, 126, 133, 146–47, 151, 157, 175, 188, 222, 230, 237, 252, 258, 260, 261

Accreditation, 1, 45, 91, 223, 263n26

Accusation, 87n41, 134, 200, 223, 232, 263n26. *See also* challenge; contestation

Adjustment, 30–32, 72, 90, 121, 141n53, 157, 203, 209, 223, 225, 233, 241, 243, 247, 257, 259, 266

Advancement: of demonstration or proof, 114, 123, 157, 204; of position or thesis, 7, 45, 91, 92n56, 105, 118, 124–25, 131, 156, 169, 237, 248, 261; of resources, 84; of results 2, 89, 101, 149, 157. *See also* emphasis

Algorithm: fuzzy, 178; genetic, 130n33, 131

Analysis: associationist, 168n24; of language, 63–70; of logical and mathematical practices, 7–10, 12–14, 17–19, 22, 24, 28, 31–39, 42, 49, 54, 58–59, 93, 95–96, 99–100, 109, 114, 123–24, 129, 137, 144, 191–93, 242, 249–50, 253, 258, 260, 265; relationist, 105; semiotic, 252; sociological, 1–3, 18, 34, 40n77, 100, 143, 206, 231–32, 262–63, 266–69; of texts, 1–4, 12, 17, 26, 45–46, 48–49, 51n110, 56, 78, 83, 89, 91–92, 101–3, 107, 112, 117, 127, 146, 161, 188, 210–11, 217, 231–33, 237, 243

Appearance: of actors, 249; as apparition, 30, 96, 98–104, 108, 117–18, 120, 123, 144, 236, 252, 254, 256, 258. *See also* disappearance; reappearance; visibility

Apprehension, as intellectual understanding, 12, 18, 19n6, 21, 35–38, 44–45, 52–54, 96–99, 103, 105, 113, 119, 123, 129, 147, 204, 232, 253, 255, 265, 268.

See also grasp, as intellectual understanding

Argument: *ad hominem*, 125; exchange of, 29, 53, 96, 146, 191, 250; informal, 27, 29, 144; logical, 26–27, 214; "strong," 138

Argumentation, study of, 29, 96, 146n69, 191

Aristotle, 44–45, 93n60, 136–37, 139, 140, 145, 193, 195

Arrangement, 5, 52, 96n69, 105–6, 113, 153, 176, 178, 208, 250, 253, 256, 261

Artificial intelligence, 1–2, 4, 9–11, 34n55, 57, 96, 118, 159–160, 162–63; fuzzy logic and, 124, 127–36, 145, 151–56, 159–60, 164, 169n26, 178n37, 175, 179, 188–91, 196–97, 201–2, 220, 223, 240

Attestation, 132, 174, 179. *See also* testimony; witness

Attribution: of claims, 200–201; to fuzzy logic, 9, 117–19, 122, 131, 136, 146, 165, 173, 194, 222, 264; of meaning, 2, 63, 83–86, 202, 205–6, 209–10, 231–32, 249, 252, 255–56; of properties, 67, 123, 144–46, 208; of status, 168, 213, 258; of truth value, 67

Autonomy: of demonstration, 73, 101, 260; of language, 60; of logical system, 91; of syntax, 21

Belmont, David (*pseud.*), 172, 178–80, 187, 189, 196, 198–204

Berenji, Hamid R., 190n1, 240n49, 240n51, 241, 242n53, 247n59

Berger, Thomas (*pseud.*), 172, 178–79, 187, 189, 198–204

Bezdek, James C., 190n1

Black, Max, 85n27, 195n14

Bloor, David, 18, 24–31, 33, 49, 88–89, 97, 107, 144, 162, 216n10, 236n40, 248n61

Bonissone, Piero P., 190n1

Bonney, Hugh, 141n54, 142n57